Major Naval Operations

Milan Vego

NAVAL WAR COLLEGE PRESS
Newport, Rhode Island

Naval War College

Newport, Rhode Island
Center for Naval Warfare Studies
Newport Paper Thirty-two
September 2008

President, Naval War College
Rear Adm. Jacob L. Shuford, USN

Provost/Dean of Academics (Acting)
William R. Spain

Dean of Naval Warfare Studies
Robert C. Rubel

Naval War College Press

Director: Dr. Carnes Lord
Managing Editor: Pelham G. Boyer

Telephone: 401.841.2236
Fax: 401.841.1071
DSN exchange: 948
E-mail: press@nwc.navy.mil
Web: www.nwc.navy.mil/press

Printed in the United States of America

The Newport Papers are extended research projects that the Director, the Dean of Naval Warfare Studies, and the President of the Naval War College consider of particular interest to policy makers, scholars, and analysts.

The views expressed in the Newport Papers are those of the authors and do not necessarily reflect the opinions of the Naval War College or the Department of the Navy.

Correspondence concerning the Newport Papers may be addressed to the Director of the Naval War College Press. To request additional copies, back copies, or subscriptions to the series, please either write the President (Code 32S), Naval War College, 686 Cushing Road, Newport, RI 02841-1207, or contact the Press staff at the telephone, fax, or e-mail addresses given.

ISSN 1544-6824

ISBN 978-1-884733-50-5

Contents

Foreword

Naval history as generally recounted is a story of battles at sea. However, it has to be admitted that since 1945 neither the United States nor any other contemporary naval power has had much of a naval history in this sense. Domination of the oceans by the United States and its allies, together with the fortunate failure of the Cold War to culminate in a test of strength between the American and Soviet navies, meant that classic naval battle gradually faded from center stage in the education and professional orientation of American naval officers. Beginning in the early years of the Cold War, the Navy became preoccupied largely with technology and the tactical proficiency that rapidly advancing naval and weapons technologies made increasingly necessary. At the extreme, of course, the advent of nuclear weapons seemed to many to leave the Navy little role in a major global conflict other than to provide invulnerable launch platforms for these weapons—and thereby a powerful deterrent that would, as it was thought, obviate their actual use. Beyond that, though, the switch to nuclear propulsion for the Navy's capital ships laid heavy technical demands on new generations of naval officers, with concomitant impact on their education and training. The result—or so contends Milan Vego in *On Major Naval Operations,* the thirty-second volume in the Naval War College Press's Newport Papers series—has been a long-standing neglect by the U.S. Navy of major naval operations and, more broadly, of the "operational" level of war or of naval "operational art."

The term "operational art" is apt to be unfamiliar to most Americans. American military officers encounter it routinely as a fixture of contemporary joint military doctrine, but even today the concept has substantially less traction within the U.S. Navy than it does in the other services. The reason is plainly that its origins are in land warfare—specifically, in large-scale land warfare as theorized by the German and (especially) Soviet militaries during the interwar period and practiced by these countries in World War II. From the latter, it migrated to the U.S. Army in the late 1970s, as the Army sought novel ways to grapple with the increasingly formidable prospect of a Soviet ground assault against Western Europe. Essentially, "operational art" refers to a level of command intermediate between the tactical and the strategic, one associated with ground command at the level of field army or corps and with the conduct of "campaigns" that unfold as a series of interconnected battles over time. That many naval officers remain unconvinced of its applicability to their own domain is not surprising, given the narrowly tactical focus of much naval warfare of the past. (Wayne Hughes's classic treatise *Fleet Tactics*, for example, begins by dismissing the utility of the concept of operational-level warfare for naval combat.) On the other hand, it is difficult to deny that naval command and control doctrine and practice today are insufficiently attentive to what in Army parlance would be called a "combined arms" approach to warfare. The tenuous relationship between the three principal naval warfare communities remains the strongest argument for a serious reconsideration by the Navy of major naval operations and operational art.

Dr. Milan Vego is a professor in the Joint Military Operations Department of the Naval War College. He has published widely on the history of German and Soviet military doctrine, and he is the author of *Operational Art* (2001) and *Joint Operational Warfare* (2008), an authoritative textbook currently utilized in the department's curriculum. In this work, he looks back to the richly instructive experience of the U.S. Navy in World War II (as well as in more recent operations during the Korean and Vietnam wars and in the Persian Gulf) in order to develop a taxonomy of naval operational art that can help inform the thinking of the Navy as a whole today.

CARNES LORD
Director, Naval War College Press
Newport, Rhode Island

Preface

Major naval operations are the principal methods by which naval forces achieve operational objectives in a conflict at sea. The U.S. Navy and other major Western navies planned and executed numerous major naval operations in World War II as part of maritime and, in several cases, land campaigns. Major naval operations have been conducted on relatively few occasions since 1945, because of the absence of a war between major naval powers. Yet the U.S. Navy has conducted several major naval operations since then. For example, in the Korean War (1950–53) UN naval forces conducted a major amphibious landing at Inchon in September 1950 (Operation CHROMITE), a major evacuation from Hungnam in December 1950, and a naval blockade of North Korea. During the Vietnam War, the U.S. Navy conducted major operations aimed to stop the flow of fighters and supplies by the sea from North Vietnam to South Vietnam in 1965–70 (Operation MARKET TIME) and conducted massive strikes and attacks by its carrier forces from "Yankee Station" in the Gulf of Tonkin. The Navy's mining of North Vietnam's ports in May 1972 (Operation POCKET MONEY) was also a major naval operation, in terms of the scale of the objective. In January 1991, in the opening phase of the Gulf War (Operation DESERT STORM), the U.S. Navy and its coalition partners conducted a series of actions that could be considered a major combined naval operation, aimed at obtaining and then maintaining control of the northern part of the Persian (Arabian) Gulf. Likewise, coalition forces conducted a series of tactical actions in the northern part of the Gulf in the opening days of the war against Iraq in March 2003 (Operation IRAQI FREEDOM).

The steady decline in the numbers of U.S. naval combatants over the past decade will most likely continue. This, in turn, will make it even more critical than it is today to employ U.S. naval forces in such a way as to maximize their enormous combat power and minimizes any advantage their opponent might have, should they be called upon to engage in combat. Combat at sea is not merely a matter of putting weapons on targets. Wars can be ultimately won only by achieving strategic or operational objectives and through selection of the proper method of employment of combat forces. It is the military objective to be accomplished that is the very heart of operational art.

In naval terms, operational art can be described as the intermediate component of naval warfare, lying between naval strategy and tactics, concerned with both the theory and the practice of planning, preparing, conducting, and sustaining major naval operations and campaigns aimed at accomplishing operational or strategic objectives at sea in a given theater. Despite rich experience in World War II, the U.S. Navy has never developed a comprehensive theory of major naval operations. This is not a trivial matter. The neglect of naval theory and the lack of sound doctrine cannot be remedied quickly, if at all, once hostilities commence.

A theory of major naval operations and maritime campaigns is the key prerequisite for writing service doctrine focused on the operational level of war. Without sound doctrine, it is hard to see how the U.S. Navy can train and prepare its operational commanders and their staffs to employ combat forces effectively. A well written and properly focused service doctrine would significantly raise the professionalism and skills of the commanders and their staffs at all levels of command. A sound service doctrine would greatly enhance operational thinking and thereby bring about badly needed balance between tactics and operational art. At the same time, the major tactical commanders—for example, carrier strike group (CSG) and expeditionary strike group (ESG) commanders—cannot successfully carry out the higher operational commander's intent without knowledge and full understanding of operational art.

This work is primarily aimed to explain and analyze major naval operations as a method of combat employment of naval forces, as well as the types of major naval operations and their planning and execution. It is divided into eight topical sections. The introduction describes the origins of major naval operations, while chapter 1 describes the meaning, purpose, prerequisites, and elements of a major naval operation. Chapter 2 explains in some detail the types of major naval operations in terms of their main purposes. Chapter 3 describes and analyzes the process of making decisions and planning major naval operations, while chapters 4 and 5 explain the key elements of an operational design and operational idea (scheme). Chapter 6 discusses the preparation for and execution of a major naval operation. The last chapter offers some thoughts on the future of major naval operations as the main method of combat employment of naval forces to accomplish operational objectives at sea.

I am greatly indebted to several individuals for their encouragement and support. The chairman of the Joint Military Operations Department at the Naval War College, Captain James Maynard, U.S. Navy, and Dr. Carnes Lord, editor of the Naval War College Press, suggested that I write a monograph on major naval operations. Their support, as well as that of the Press staff, was critical for the completion of the project. I am also very thankful for a superb and patient job to Albert F. Fassbender, who proofread my manuscript. Finally, I want to thank the Naval War College as an institution for providing support and resources for the publication of this work.

List of Figures

Introduction

In the era of sail and until the late nineteenth century, the principal method of combat employment of one's fleet to attain an operational and sometimes strategic objective was a "decisive naval battle." Some of the "decisive battles"—for example, the battle of Trafalgar in October 1805—led to a drastic change in the situation at sea. The methods of combat employment of naval forces gradually changed because of the effects of the new technological advances in the middle and late nineteenth century. Great advances in steam propulsion and the invention of the internal combustion engine made it possible to fit powerful engines on even small ships. The advent of torpedoes and mines led to the design of new, small platforms capable of posing a serious threat to larger ships. This in turn led to a proliferation of small warships of all types and classes. The numerical strength of the major navies steadily increased. In addition to battleships and cruisers, they also included a large number of smaller surface combatants, such as destroyers, torpedo craft, gunboats, and auxiliaries.[1]

The importance of a decisive battle dropped steadily with the increase in size of the fleets of major navies and corresponding changes in their compositions. Instead of a single decisive battle to achieve an operational and even strategic objective, war at sea between two strong opponents was fought over a large area and almost continuously. Numerous tactical actions were conducted by both large and small surface combatants. The deployment of one's fleet forces became an integral part of a major clash with the opposing fleet forces.

The Russo-Japanese War of 1904–1905 included a series of major and minor naval actions, all related in terms of their main purpose, place, and time. These actions took place over large parts of the Yellow Sea, the Sea of Japan, and even part of the Pacific Ocean.[2] In retrospect, the battle of Tsushima in May 1905 was the last "decisive" naval battle in history. Yet this fact was not recognized by the theoreticians and practitioners of the day. Up to the beginning of World War I, all major navies of the day planned to fight a decisive battle (also called "general fleet action") aimed at obtaining command of the sea in a single clash of battle fleets.

In the last decade before the outbreak of World War I, the major navies continued to build both large and small naval vessels in large numbers. For example, by 1914, the Royal Navy had in service 542 warships, including 68 battleships, 110 cruisers, and 218 destroyers. The Imperial German Navy then had 301 ships (37 battleships, 48 cruisers, 142 destroyers, 47 torpedo boats, and 27 submarines).[3]

The advent of the submarine and aircraft added second and third dimensions to the maritime battlefield. The ability to control the movements and actions of the fleet's elements over large ocean areas was considerably enhanced by the invention of new signaling devices and of wireless radio. Fleet commanders were able to communicate with warships many hundreds of miles out at sea. The cumulative effect of all these changes was that major navies of the day were capable of conducting actions almost continuously, over large sea/ocean areas, by employing diverse platforms and weapons. The very size of the major navies of the day, with their widely dispersed home bases and installations, made it increasingly difficult, if not impossible, to achieve decisive results just by fighting a single general fleet action.

During World War I several large-scale fleet-versus-fleet actions took place in the North, Mediterranean, Adriatic, Baltic, and Black seas. For example, the battle of Jutland, fought on 31 May–1 June 1916, came closest to what is known as a "major naval operation." It comprised a series of small engagements and attacks aimed to accomplish an operational objective. The German plan was to bring the strength of the British Grand Fleet down to parity with that of the High Seas Fleet.[4] The British operational objective was the destruction of the High Seas Fleet. Both sides also conducted a series of actions to provide combat support to their battle fleets.[5]

The battle of Jutland was won tactically by the High Seas Fleet, in terms of losses in materiel and personnel inflicted on the Grand Fleet. The British lost fourteen ships (three battle cruisers, three armored cruisers, eight destroyers and torpedo boats) and some 6,100 men (out of sixty thousand total), while the German losses amounted to eleven ships (one predreadnought battleship, one battle cruiser, four light cruisers, and five destroyers/torpedo boats) and about 2,550 men (out of thirty-six thousand total).[6] However, the operational victory was clearly won by the British, because in the aftermath of the battle the situation in the North Sea and adjacent sea areas remained essentially what it had been before the battle took place. Both opponents continued to watch each other and behaved essentially as active fleets-in-being. The Entente's blockade of the Central Powers was not weakened. The Royal Navy continued to ferry troops and supplies to France. The Germans retained their naval control of the Baltic theater.

The first major naval operation against the enemy coast was conducted by the Austro-Hungarian fleet, shortly after Italy's decision to enter the war on the side of the Entente powers on the night of 23/24 May 1915. This operation had been planned by the Austro-Hungarian navy in 1910, even though Italy was formally the country's ally at that time. The main objective of that operation was to disrupt and significantly delay the movement and transport of mobilized Italian troops by rail along the Adriatic's western coast. Another objective was to create fear and possibly panic among the Italian populace in the coastal area.[7]

During World War I, the first major joint/naval operations emerged; the Entente's amphibious landing at Gallipoli in April 1915 and the German landing in the Gulf of Riga in October 1917 (Operation ALBON) are the best-known examples. The principal objectives of the Gallipoli landing operation were to take Turkey out of the war, open a direct link with the Entente's embattled ally in Russia, force the Germans to shift troops from the Russian front, and influence Greece to side openly with the Entente powers.[8] The landing at Gallipoli eventually failed. The last British troops were evacuated in January 1916. The losses on both sides were heavy. The Entente committed about 490,000 troops to the operation and suffered 141,100 casualties (including about 44,100 killed). The Turks employed some 500,000 troops and had about 251,300 casualties (including some 86,700 killed).[9]

A more successful amphibious landing operation was conducted by the German navy and the army in the Gulf of Riga. This was the first time the Germans had planned and executed a joint operation of such size and complexity. The operational objective of the landing was to open the Gulf of Riga and thereby threaten the rear of the Russian Twelfth Army, defending the Baltic coast. The Germans achieved complete surprise, and the entire operation was a resounding success. The Twelfth Army was eventually destroyed, and the threat to the flank of General Oskar von Hutier's Eighth Army was eliminated.[10]

World War I at sea proved that the fleets of the major opponents were too large and deployed too widely to be destroyed during a single battle, or even two. It signaled the final demise of the decisive battle, the general fleet action, and demonstrated that *operational* objectives in the theater could be accomplished only by a series of related naval battles and engagements, sequenced and synchronized in time and place—that is, by a major naval operation, in modern terms. Deployment, clash of opposing forces, pursuit, and withdrawal/redeployment were meshed to constitute a seamless whole. The entire naval operation was planned, prepared, and conducted by a single commander. It was based on a certain idea and a common plan.

In the interwar years, most major Western navies and also the Japanese navy focused almost exclusively on the practical application of operational warfare through planning, war gaming, and exercises at sea. In 1927, the U.S. Naval War College adopted for the first time the study of "operational" problems in addition to "strategic" and "tactical" problems. This practice continued in the 1930s. Also, in the war games held at the Naval War College in the 1930s, the U.S. Navy repeatedly tested its Plan ORANGE for a hypothetical war with Japan. In the early 1930s, the U.S. Marine Corps embarked on an intensive effort to create its first operational doctrine for conducting large-scale amphibious landings. The result was the *Tentative Manual for Landing Operations*, issued in 1934. This manual borrowed heavily from the lessons learned in the Entente's unsuccessful Gallipoli landing in 1915 and Germany's successful Operation ALBION. After 1935 this document was used as a guide for future amphibious exercises and for research and development.[11]

The German navy used planning games, war games, and naval exercises extensively in preparing for a future war at sea. By the early 1930s the German navy had introduced "operational," in addition to strategic and tactical, war games. In the late 1930s, as a result of these games, the German navy's high command became convinced of the need to seize parts of the southern Norwegian coast and the French Atlantic coast to escape the constraints imposed by the geography of the North Sea and to employ its forces operationally in the Atlantic. The tactical concept of using U-boats at night and on the surface was first described in 1922, in a study written by two German naval officers. Their ideas were refined during war games in the early 1930s and became part of doctrine in 1935.[12] This innovative concept was applied for the first time at the operational level in German naval exercises early in 1939.

In contrast, the Soviets focused on developing a theory of naval operations as part of their overall approach in developing operational art. Their theories were incorporated into the Red Navy's doctrine. The Soviet *Field Manual of 1930* (BU-30) was the first doctrinal document to spell out the rudiments of joint operational employment of naval forces and ground troops. Afterward, the Soviets developed the theory of what they called "naval operations"—specifically, reconnaissance, amphibious landings and anti-amphibious operations, and operations on sea communications. Five years later, the Soviet navy adopted the new *Combat Manual of Naval Forces* (BU-MS-37). For the first time, the Soviets elaborated "naval operations" and "day-to-day (routine) activities" as the principal methods by which the combat employment of naval forces and aviation would accomplish operational objectives. The new manual envisaged about a dozen types of naval operations (operations aimed at destroying the enemy forces at sea, operations against coastal "objects" [installations], operations on sea communications and blockade, operations in support of own sea communications and counterblockade, landing operations, antilanding operations, and operations in support of the army flank). The day-to-day activities of the fleet forces were aimed at accomplishing operational objectives by obtaining and maintaining what the Soviets called a "favorable operational regime" in one's own coastal waters and deployment area.[13] The Soviet views on the character of modern warfare at sea and operational art were incorporated into the *Provisional Manual on Conduct of Operations* (NMO-40), issued in 1940. However, the Soviet theory of naval operations was very poorly applied during the country's war with Nazi Germany.

In World War II, all the major navies conducted, independently and in cooperation with other services of the armed forces, a large number of major naval operations in all the maritime theaters of war. Among many major naval operations, a few stand out because of their importance to the course of the war. Major clashes of opposing surface forces in the Pacific, notably the battles of the Coral Sea in May 1942, Midway in June 1942, the Philippine Sea in June 1944, and Leyte in October 1944, were not "battles" but major naval operations. In the Atlantic Ocean, the Germans conducted a major naval operation (RHEINUEBUNG) with "Combat Group *Bismarck*" (the 42,600-ton battleship *Bismarck* and the heavy cruiser *Prinz Eugen*) aimed at interrupting Allied maritime traffic in the North Atlantic in May 1941. The escape of two German battle cruisers *(Scharnhorst, Gneisenau)* and one heavy cruiser *(Prinz Eugen)* from Brest to Kiel through the English Channel (Operation CERBERUS) in February 1942 was also a major naval operation. The purpose was to redeploy these heavy ships to Norway from threatening Allied maritime traffic in southwest approaches to England and thereby strengthen German defenses against possible Allied invasion. The British carrier attack on the Italian naval base at Tarent in November 1940 (Operation JUDGMENT) and the Japanese attack on Pearl Harbor in December 1941 (Operation HAWAII) are also examples of major naval operations aimed to destroy a major part of an enemy fleet at its base. The Allies also conducted many major amphibious landing operations in all theaters, but especially in the Pacific, during World War II.

Since the end of World War II most regional wars have not involved major navies on both sides, and so only a few major naval operations have been conducted. However, the blockade of North Korea's coast during the Korean War (1950–53) and the U.S. blockade of South Vietnam's coast (Operation MARKET TIME, March 1965–December 1972) are examples of major combined naval operations. Also, the Israelis planned and conducted what can be considered a major naval operation to obtain local sea control off the Egyptian coast in the Mediterranean and off the Syrian coast during the Yom Kippur/Ramadan War of October 1973. The British recapture of the Falklands in 1982 (Operation CORPORATE) was a major naval/joint operation with a limited strategic objective. Relatively recently, the operations of the coalition naval forces in the Gulf War of 1990–91 to obtain and maintain control of the northern part of the Arabian (Persian) Gulf constituted a major combined naval operation.

Notes

1. In 1900, the Royal Navy consisted of 205 warships: forty-five battleships, 126 cruisers, and thirty-four torpedo gunboats; the French navy had in service thirty-three battleships, thirty-eight cruisers, and twenty-one torpedo gunboats; Ronald B. St. John, "European Naval Expansion and Mahan, 1889–1906," *Naval War College Review* (March 1971), p. 74.
2. Adolf Feil, ed., *Grundlagen der Theorie der operativen kunst der SSK,* trans. Dieter Foerster and Bernd Kulbe (Dresden: Friedrich Engels Military Academy, March 1989; translation of *Operativnoye iskusstvo voyenno-morskovo flota*), p. 11.
3. St. John, "European Naval Expansion and Mahan, 1889–1906," p. 74.
4. Elmer B. Potter and Chester W. Nimitz, eds., *Seemacht. Eine Seekriegsgeschichte von der Antike bis zur Gegenwart,* rev. ed. (Hersching: Manfred Pawlak, 1986; translation of *Sea Power: A Naval History* [Englewood Cliffs, N.J.: Prentice Hall, 1960]), p. 358.
5. Feil, ed., *Grundlagen der Theorie der operativen kunst der SSK,* p. 11.
6. Geoffrey Bennett, *Naval Battles of the First World War* (New York: Scribner's, 1968), p. 242.
7. Potter and Nimitz, eds., *Seemacht,* p. 433.
8. James B. Agnew, "From Where Did Our Amphibious Doctrine Come?" *Marine Corps Gazette* (August 1979), p. 53.
9. Alan Moorehead, *Gallipoli* (London: Hamish Hamilton, 1956, 1967), p. 361.
10. Agnew, "From Where Did Our Amphibious Doctrine Come?" p. 56; Carl O. Schuster, "Baltic Assault: Operation Albion," *Command Magazine* (July–August 1993), pp. 56–57; Randolf Kugler, *Das Landungswesen in Deutschland seit 1900* (Berlin: Oberbaum Verlag, 1989), p. 35.
11. Agnew, "From Where Did Our Amphibious Doctrine Come?" pp. 58–59; Allan R. Millet, "Assault from the Sea: The Development of Amphibious Warfare between the Wars—The American, British, and Japanese Experiences," in *Military Innovation in the Interwar Period,* ed. Williamson Murray and Allan R. Millet (Cambridge, U.K.: Cambridge Univ. Press, 1998), p. 75.
12. Holger Herwig, "Innovation Ignored: The Submarine Problem; Germany, Britain, and the United States, 1919–1939," in *Military Innovation in the Interwar Period,* ed. Murray and Millet, p. 234.
13. The manual was written by a group of instructors of the naval academy, led by Captain First Rank S. Stavitskiy. The final version of the manual was written by a commission headed by Captain First Rank I. S. Isakov; A. Gakkel, "The Soviet Art of Naval Warfare in the Period between the Wars (1921–1941)," *Morskoy Sbornik* 10 (October 1976), p. 15; A. V. Basov, *Flot v Velikoy Otechestvennoy Voyne 1941–1945. Opyti operativno-strategicheskogo primeneniya* (Moscow: Nauka, 1980), pp. 50–52.

CHAPTER ONE

Major Naval Operations

There is no common agreement on a definition of what constitutes a major naval operation. The term "naval operation," so often used by the U.S. Navy and other Western navies, is too broad and imprecise. For example, "naval operation" in the United States is characterized as pertaining to "a naval action or the performance of naval missions, which may be strategical, tactical, logistical, or training." This definition clearly does not make any distinction between the objectives to be accomplished at sea and corresponding methods of combat force employment. It confuses the issue by mixing the combat employment of naval forces with logistics and training. An alternative meaning of the same term is given as "the process of carrying out or training for naval combat to gain the objective of any battle or campaign."[1] The fact that the phrase "major naval operations" does not appear in either definition is clear proof of the complete absence of operational art, which is, incidentally, the focus of all joint doctrinal documents in the U.S. military.

In comparison, the former Soviet navy and Russian navy devoted extraordinary efforts to developing a theory of major naval operations in the early 1930s and through the 1980s. Naval operations constituted the very heart of their naval operational art. Yet the Soviets were distinctly unsuccessful in applying their theories in practice during the war with Nazi Germany, 1941–45. From 1945 until the collapse of the Soviet Union in 1991, the Soviet navy was never involved in real combat. Hence, one can only speculate whether its actual performance would have been any better then than it was during World War II.

In one of the many Soviet definitions of the term, a naval operation *(morskaya operatsiya)* was described as a series of naval battles, engagements, and strikes coordinated in terms of objective, place, and time and conducted in an assigned area (zone) of an oceanic or sea theater of war. It is carried out by specially established groups, independently or in cooperation with formations, forces, or troop units of other services, according to a common idea and plan for accomplishing a single, especially important operational or operational-tactical task. Naval operations are controlled by the commander of a fleet.[2]

In generic terms, a *major naval operation consists of a series of related major and minor naval tactical actions conducted by diverse naval forces and combat arms of other services, in terms of time and place, to accomplish an operational (and sometimes strategic) objective in a given maritime theater of operations. Major naval operations are planned and conducted in accordance with an operational idea (scheme) and common plan. They are normally an integral part of a maritime or land campaign, but they can sometimes be conducted outside of the framework of a campaign.*

Tactical actions in a major naval operation can be fought on the surface, beneath the surface, in the air, and in some cases on the coast. They are not just a simple arithmetic sum, but they are all related and conducted within a given operational framework. A tactical action fought outside of the framework of a major naval operation would generally not contribute to the accomplishment of the assigned operational objective and therefore would represent a waste of resources and time.

Naval tactical actions can range from actions in which weapons are not used, such as patrols and surveillance, to attacks, strikes, raids, engagements, and naval battles. As the term implies, they are aimed at accomplishing tactical objectives in a given part of a maritime theater. In some cases, a series of diverse tactical actions conducted over time can lead to the accomplishment of an operational objective.

In the past, a *naval battle* was the main method of accomplishing a major tactical objective as a part of a major naval operation. It consisted of a series of related attacks, counterattacks, strikes, and counterstrikes coordinated in time and place. It was characterized by relatively long duration—several hours or even longer.[3] A naval battle was conducted with the participation of several naval combat arms, and often the combat arms of other services (e.g., air forces or troops defending a coast).

A decisive defeat of the enemy fleet in a naval battle can sometimes result not only in an operational victory but also in a strategic victory. To complicate the matter, a naval battle can be tactically won but yet represent an operational failure. For example, the battle of the Coral Sea in May 1942 (Operation MO, for the Japanese) was tactically a draw, but there is no doubt that the Allies achieved their operational objective, in that the Japanese abandoned their attempt to seize Port Moresby by assault from the sea.[4]

In the past, a *naval engagement* consisted of a series of related strikes/counterstrikes and attacks/counterattacks aimed to accomplish the most important tactical objective in a naval battle. For example, in the battles of the Coral Sea in May 1942, Midway in

June 1942, and the Philippine Sea in June 1944, the clashes of the opposing carrier forces were naval engagements.

With the advent of missiles and other long-range, highly precise, and lethal weapons it became possible to destroy the enemy force at sea or on the coast at much longer range than by using guns or torpedoes. Hence, a new method of combat force employment called "strike" gradually emergerd as the principal method of accomplishing major tactical and sometimes operational objectives in war at sea and in the air. A well prepared strike is difficult to repulse.[5]

Depending on the target to be destroyed or neutralized, one may differentiate tactical, operational, and strategic strikes. A strike can be conducted by a small number of platforms of a single type of force—for example, missile craft, submarines, or attack aircraft (helicopters). On an archipelagic coast, missile- or torpedo-armed surface combatants can conduct strikes from ambush against a much stronger hostile force. A strike can be carried out by using conventional weapons (missiles, torpedoes, guns, bombs, etc.) or weapons of mass destruction (WMD). In a strike conducted with several weapons, longer-range weapons are generally used first, to weaken the enemy's defenses; short-range weapons then finish off the target.

A broader form of strike is a *naval raid*—conducted by a single or several naval combat arms to accomplish a tactical objective as a part of a major offensive or defensive naval operation. The aim is usually to deny temporarily some position or to capture or destroy an enemy force, coastal installation, or facility.[6] Temporary or local control of the sea is not a prerequisite for the success of these actions. The stronger fleet can also conduct raids to divert the enemy's attention or to force the enemy to react in a secondary sector.

A naval raid is usually conducted against an objective that the enemy considers so valuable that its loss or serious degradation cannot be ignored. A larger purpose of a naval raid is to accomplish some temporary advantage, with the threat of future repetition. Besides destroying installations, facilities, or enemy forces on the coast, a naval raid can enhance one's morale or diminish the enemy's. It can range in scale from an action of a very small force to a large tactical action; it can have many features of a major naval operation. The effect of a raid, like that of surprise, is usually transitory. However, depending on the enemy's reaction, the consequences of a well executed raid can be much greater than initially planned.

The most frequently conducted tactical action using weapons is a *naval attack,* a combination of tactical maneuver and weapons used to accomplish a minor tactical objective. A naval attack can be conducted independently or as part of a strike or raid. A naval attack can be conducted by a single or several types of platforms. It is intended to destroy or neutralize a single enemy platform, tactical groups, or targets ashore. Attacks can be distinguished by the type of weapons used—missile, gun, torpedo, bomb, depth charge, or a combination thereof.[7] The success of an attack depends largely on the surprise achieved, the degree of skill in maneuvering the platform (or groups) to obtain positional advantage, and the range, lethality, and precision of the weapons used. The simplest example of a naval attack is that of sinking a single platform or damaging a naval vessel or merchant ship.

A major naval operation is conducted as an integral part of a maritime, sometimes land, campaign. It is aimed to bring about a drastic change in the operational situation in a given part of the maritime theater. However, if the major naval operation is only partially successful, the operational situation is likely to remain as it was prior to the operation's execution. A major naval operation can also have a strategic effect, though its main objective is operational in scale. For example, the operational objective of the Japanese surprise attack on Pearl Harbor on 7 December 1941 was to destroy the major part of the U.S. Pacific Fleet, in order to prevent it from interfering with the invasion of the Philippines. However, the unintended and highly negative effect for Japan was that U.S. public opinion shifted from isolationism to unequivocal support for the war until complete victory over Japan.

A maritime campaign is predominantly fought on the open ocean and in sea areas adjacent to a continental landmass. One's naval forces would play the most important role in such a campaign. A land campaign in the littoral area would also involve participation of naval forces. In general, the closer to the continental landmass or large archipelagoes one's naval forces have to operate, the more the success of their actions will depend on close cooperation with other services or a high degree of jointness.

The key to ultimate success in war at sea is avoiding situations in which one's objectives are accomplished predominantly—or even worse, exclusively—through force-on-force encounters or attrition. Attrition warfare not only results in much higher losses, even for the stronger side, in terms of human lives and materiel but also is inherently protracted. One's forces are tied down as long as a given operational objective remains unaccomplished. Therefore, they cannot be employed for other urgent tasks in another ocean or sea area. This happened to the Allies in the struggle for Guadalcanal between August 1942 and February 1943. The initial major naval operation—the amphibious landing on Guadalcanal (Operation WATCHTOWER) on 7 August 1942—was highly successful. Afterward, however, the Allies became progressively involved in a series of small but costly actions with Japanese forces on land, at sea, and in the air. This attrition phase lasted for almost seven months before the Japanese decided to give up their attempts to regain control of Guadalcanal. By then the protracted struggle had considerably slowed the Allied operational tempo; no further major landings up the Solomons chain had been possible until Guadalcanal was secure. A good argument can be made that had the Allies conducted a successive major naval operation, the struggle for Guadalcanal would have ended much earlier and with far less Allied loss in ships, aircraft, and personnel.

The best and most proven way of avoiding attrition at the operational or strategic level is to plan and execute major operations or campaigns, respectively. By conducting a major naval operation, the stronger side at sea can defeat the weaker side in an ocean/sea area and within the time frame of its own choosing. Major naval operations are normally planned and conducted when decisive results have to be accomplished in the shortest time possible and with the least losses for one's forces.[8] Successful major naval/joint operations can also contribute considerably to shortening a war at sea.

Major naval operations are the most effective way to avoid attrition warfare at the operational level of war. They are especially critical for success in the initial phase of a war, when the stronger side needs to obtain and then maintain a desired degree of sea control in a certain part of a maritime theater. This can be accomplished by destroying or seriously weakening a major part of the enemy fleet at sea or of its bases, by destroying a major enemy convoy or protecting one's own, or by seizing or defending a major island or choke point. Major naval operations require usually much larger forces than do single battles or engagements. They are planned and conducted in accordance with the tenets of operational art.

Purpose

In terms of principal purpose, a major naval operation can be offensive or defensive (see figure 1). An offensive major naval operation is normally planned by the stronger side at sea, but it also can be planned by the side on the defensive. Normally, the stronger side at sea would mount a single major naval operation, or several in succession, to obtain and then maintain sea control in a specific part of a maritime theater. Such operations can also be aimed to reduce or eliminate the threat posed by a numerically larger force in order to facilitate operations in another part of the maritime theater, as the Japanese surprise attack on Pearl Harbor illustrates. An offensive major naval operation can also be planned as part of a defensive campaign, an example being the Japanese decision to commit a major part of its Combined Fleet to the defense of the Philippines in October 1944. Major naval operations against enemy maritime trade and amphibious landing operations are inherently offensive.

A major naval operation with an offensive purpose is usually planned and conducted when one's forces possess at least temporary sea control in a given maritime area of

FIGURE 1
Types of Major Naval Operations

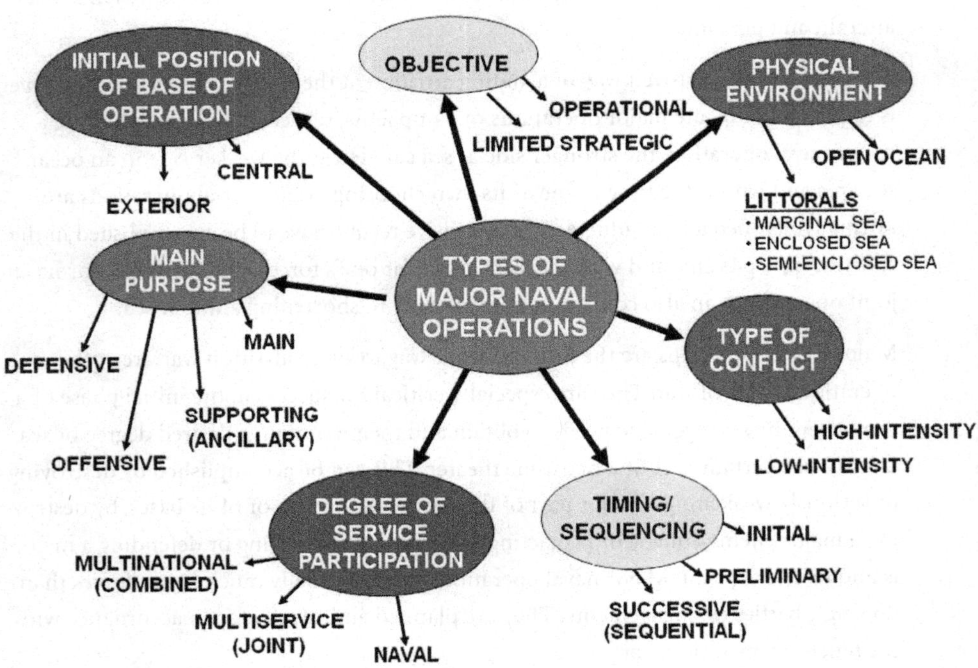

operations. Yet the fact that the enemy possesses control of the sea and air does not necessarily prevent the inferior side from mounting a major offensive naval operation. For example, in the Pacific during World War II most Allied amphibious landings were conducted after obtaining merely temporary and local sea control. In contrast, major naval operations in enclosed and semi-enclosed seas can be conducted within either strategically offensive or strategically defensive frameworks. For either one, limited and temporary sea control should be obtained first. Major naval operations aimed at protecting maritime trade can be conducted even when command of the sea is in dispute. The main prerequisite for success then is to ensure at least control of the sea, the subsurface, and the air in the proximity of a large convoy. For example, the Royal Navy mounted several major operations in 1941–42 (Operations HARPOON/VIGOROUS, PEDESTAL, etc.) to supply the besieged island of Malta, although Axis forces possessed overwhelming strength in the air. It is also possible to conduct a major naval operation when friendly forces control only the air, as the German navy did in the invasion of Norway in April–June 1940.

A defensive major naval operation is usually planned and executed when one side is forced onto the defensive at sea because of permanent or temporary weakness in a certain part of the maritime theater. The weaker side at sea will conduct major defensive naval operations against enemy attacks on its naval bases or anchorages and against amphibious landings, attacks on its coastal installations/facilities, and major evacuations of friendly forces or civilians. Both sides in a war at sea will occasionally mount major naval/joint operations in defense of maritime trade.

Based on the degree of participation of naval forces and other services, naval, joint (multiservice), or combined (multinational) major naval operations can be differentiated. A major naval operation is conducted predominantly by the navy, although air or even ground forces can take part as well. Examples of major operations conducted by naval forces are the battles of Cape Matapan (March 1941), the Coral Sea (May 1942), Midway (June 1942), and the Philippine Sea (June 1944).

A major naval/joint operation is planned and conducted by forces of the navy but with the involvement of one or more other services. In a maritime theater encompassing a large ocean or sea area, major naval operations would be conducted with the significant participation of the air force. Major naval operations in littoral waters are likely to be conducted with the participation of all three services. All major amphibious landing operations are inherently joint, as are major operations against the enemy's maritime trade or in defense of maritime trade in the littorals.

A major naval/combined operation is conducted with two or more navies or services of the armed forces. A major amphibious landing operation is sometimes combined, as were the invasion of Sicily (Operation HUSKY) in July 1943, the landing in Normandy (NEPTUNE) in June 1944, and the UN amphibious landing at Inchon (Operation CHROMITE) in September 1950. Today, combined major operations constitute a frequent method of employing naval forces in low-intensity conflicts. In fact, because of the downsizing of most navies, such major naval operations might well become routine for major regional contingencies or general conflict.

Sometimes several major naval operations take place in a single maritime theater of operations. Such a situation would occur in the initial phase of a new campaign when forces must accomplish several operational objectives sequentially or simultaneously. In such a case the major naval/joint operation aimed to achieve the most important operational objective will be the main or principal operation, and the others will be supporting (ancillary) operations. For example, the Allied amphibious landing at Leyte on 20 October 1944 was the main or principal major naval/combined operation. The actions of Task Force 38 (TF 38) to provide operational cover and support between 17 October and 27 November constituted a supporting major naval operation.

In terms of their timing, initial, successive, and preliminary major naval operations can be differentiated. The initial major naval operation is planned and conducted to accomplish the initial and the most operational objective in a campaign. For example, the Allied invasion of the Gilberts in November 1943 (Operation GALVANIC) was the initial major naval operation in a series of major naval operations within the Central Pacific campaign. A successive (or consecutive) major naval operation normally starts during or shortly after the initial major naval operation. Its objective is to consolidate or expand the operational success accomplished in the preceding major operations. The U.S. invasions of the Marshalls (Operations FLINTLOCK and CATCHPOLE) and the Marianas (FORAGER) are examples of successive major naval operations. TF 38's series of air strikes and other actions against the Ryukyus, Formosa, and Luzon between 10 and 17 October 1944 constituted a preliminary major operation for the invasion of Leyte.

In the past, most major naval operations have been conducted relatively close to the coasts of the continents, large islands, or oceanic archipelagoes. A large number of major naval operations have taken place in enclosed or semi-enclosed seas (collectively called "narrow seas"). Today, because of enormous advances in the range and lethality of weapons, even coastal navies composed of small surface ships, submarines, and land-based aircraft can conduct major naval operations. Such operations are generally more complicated to prepare and execute than those on the open ocean, because they involve diverse forces and assets of all services of the country's armed forces. They also differ from major naval operations on the open ocean because of the much smaller operational space and employment of smaller combat forces.

Operations on Interior versus Exterior Lines

A major naval operation can be conducted along interior or exterior lines of operations. Interior lines are those within an area completely or partially enveloped by the enemy. Hence, they originate at a central position.[9] Admiral Alfred T. Mahan (1840–1914) observed that interior lines represent in fact a central position prolonged in one or more directions. They favor sustained interposition between separated bodies of an enemy, with a consequent opportunity to concentrate against one while holding the other in check, with a force that is possibly inferior. "Interior lines" implies that from a central position one can assemble more quickly on either of two opposite fronts than the enemy can and therefore can utilize forces more effectively.[10]

Moving along *interior* lines, one's naval force can be interposed between two or more parts of the enemy force. A major part of one's force should then concentrate as quickly as possible in the sector of main effort to destroy the bulk of the enemy force. Interior lines are inherently shorter than those the hostile force uses, moving on the periphery. They can be used effectively in both offense and defense. The key prerequisites for the success of major naval operations on interior lines are sufficient space and ability to move forces quickly.[11] For example, in the Russo-Japanese War of 1904–1905, the Japanese fleet occupied a central position between the Russian major naval bases at Port Arthur and Vladivostok and checkmated both places. The Japanese armies could be transported across the Tsushima Strait and through the Yellow Sea, because the Japanese fleet was interposed between the Russian Far Eastern Squadron in Vladivostok and the force based in Port Arthur. Thus, the Japanese fleet commander, Admiral Heihachiro Togo (1846–1934), could prevent the Russian squadrons from joining in order to interfere with Japanese communications. In another example, U.S. Navy task forces 58 and 38 operated along interior lines during the battle of the Philippine Sea and the battle for Leyte, respectively.

A force is said to operate along *exterior lines* when its lines of movement are separated by those of the enemy. These lines are generally longer than the shortest line the enemy force can use. A force moving along exterior lines can select the point of attack along the enemy's periphery. The most important prerequisite for the proper use of exterior lines is that each part of one's force possess sufficient combat strength.[12]

The advantages and disadvantages of exterior lines are the reverse of those of interior lines. Exterior lines allow simultaneous concentric actions from several directions against the enemy's center.[13] They generally facilitate shifting forces to meet external threats while maintaining communications and covering the approaches to the enemy's force. The most inherent advantage of a force moving along exterior lines is that it can threaten the enemy with envelopment. However, such a force moves along longer lines of operations than its opponent does. The greater the distance between the base of operations and the attack objective, the longer the lines of supply. Also, unless each force element is stronger than the enemy force opposed to it, there is a constant danger that one's force will be defeated in detail.

Not only must a force operating along exterior lines move faster than an opponent moving along interior lines, but its adjacent elements have to keep within mutual supporting distance to avoid being defeated in detail. Such a major naval operation also requires precise synchronization of the movement and actions of forces. This is difficult to achieve, especially when the distance between one's base of operations and the physical objective is great. In the Leyte operation, the Japanese First and Second diversionary attack forces and the Mobile Force (main body) started their movements toward the Philippines from exterior positions. Likewise, the U.S. TF 38 carrier force operated from the exterior position in regard to its targets on Luzon and the central Philippines and adjacent sea areas.

The Allied amphibious force that landed at Leyte also operated from the exterior position during its transit and approach phases.

Characteristics

The main characteristics of a modern major naval operation are the complexity of its planning, preparation, and execution; participation of diverse naval combat arms and often combat arms or branches of other services; diversity of tactical actions taking place in a relatively large part of the theater; high intensity of combat and losses; extensive use of electronic warfare techniques; and highly complex combat support.

Because of the enormous increase in the combat potential of platforms, major naval operations of the future are likely to be highly intense. This will be especially true in a littoral environment, where the combination of small area and short distances can result in more intense employment of diverse platforms and weapons. For that reason, in major naval operations today high consumption rates of fuel and ammunition should be expected.

Major naval operations differ considerably from naval tactical actions in terms of force composition/mix, the size of the sea/ocean area in which combat takes place, the duration, and the scale of logistical support and sustainment. They normally require larger and more diverse forces than naval tactical actions. For example, in the Midway operation the Japanese Combined Fleet employed four separate groups with 165 warships.[14] The U.S. force that took part in that battle consisted of seventy-six warships and about 110 land-based Army, Navy, and Marine aircraft based on the island of Midway.[15] Some major naval operations conducted in World War II involved fewer ships and aircraft. For example, in the "*Bismarck* Chase" (RHEINUEBUNG for the Germans) the Germans employed the *Bismarck* combat group, six escort tankers, and one supply ship. Additionally, three destroyers, two patrol ships, two blockade runners *(Sperrbrecher)*, one minesweeper flotilla, thirteen U-boats, one Italian submarine, and three Luftwaffe air groups were used in support.[16] The Royal Navy employed directly or indirectly five battleships, three battle cruisers, two aircraft carriers, eleven cruisers, thirty-three destroyers, and eight submarines to defeat Combat Group *Bismarck*.[17]

Enormous increases in force mobility and in the range and lethality of modern weapons allow operational objectives to be accomplished with smaller forces than would have been possible only a few decades ago. Today's major blue-water navies possess far fewer ships than they did during World War II or the early 1950s. Almost any major naval operation designed to enforce United Nations Security Council sanctions would be a multinational effort, as the coalition's enforcement of UN sanctions against Iraq in 1990–91 illustrates. Likewise, major naval operations in littoral waters will include more small ships and land-based aircraft than would operations conducted on the open ocean.[18]

A major naval operation requires the employment of diverse naval combat arms and often the combat arms of air forces. In the littoral waters, such operations would also involve combat arms/branches of ground forces.

Operational deployment of forces taking part in a major naval operation on the open ocean takes place over a large part of a maritime theater. The majority of tactical actions in major operations are conducted in a specific maritime area of operations. For example, the forces that took part in the Midway operation were deployed between the Japanese home islands and the U.S. west coast, and from the Aleutians to the Central Pacific. The Leyte operation was fought in an area covering about 432,000 square miles (including eighty-five thousand square miles of the Philippines) and encompassing a major part of the western Pacific and parts of the Central, South, and southwestern Pacific. The water area in which the actual fighting occurred, however, encompassed only about 110,000 square miles.[19]

Major naval operations in the future are most likely to take place in enclosed or semi-enclosed seas, such as the Persian (Arabian) Gulf, the Yellow Sea, or such peripheral seas as the East China Sea. Yet a blue-water navy like the U.S. Navy would deploy its forces, other than those forwardly deployed, from basing areas (for instance, in the continental United States) or other maritime theaters.

A major naval operation lasts considerably longer than a naval tactical action. For example, a naval battle or engagement is fought for several hours or less, but a major naval operation can last a week or more. In the case of a low-intensity conflict or peace operations, major naval operations can extend for many months. In general, a major naval operation lasts longer if the distance between the base of operations and the operating area is very great or if the opposing force is difficult to destroy or neutralize because of its size or peculiarities of the physical environment. Such an operation also lasts much longer if the objective is to a significant extent nonmilitary in its nature, as in support to counterinsurgency or peace enforcement.

A major naval operation conducted in the littorals differs considerably from one on the open ocean, because of the much more complex physical environment and short distances. Among other things, geomorphological and hydrographic (or oceanographic) features in an enclosed or semi-enclosed sea greatly affect the employment of naval platforms and their weapons and sensors. The coast in a typical narrow sea is usually highly indented and fronted by a large number of islands and islets. This in turn greatly restricts the maneuverability of one's surface ships, especially major surface combatants and submarines. In shallow waters, large surface ships have to reduce their speeds greatly. Navigational conditions in narrow seas are often made difficult by the presence of shoals, reefs, strong tides, and currents.

A major naval operation in a typical narrow sea will most likely encompass the entire operating area. The combat will take place on the surface and the subsurface, in the air, and on the coast. Diverse combat arms of the navy and other services will be extensively employed. Combat actions will be short and intense and will probably result in high losses.

One of the main features of modern naval combat in a typical narrow sea will be quick and frequent changes in the operational or even strategic situation. Modern electronic warfare techniques will be extensively used by both sides. This will cause great difficulties in the use of sensors and guided weapons or even make their use impossible. The high speed of modern ships and aircraft and their ability to combine maneuver with fires make it possible to achieve surprise as well as to gain superiority in place and time.

Combat actions in an enclosed or semi-enclosed sea will occur largely at night or in bad visibility. Because of the limited visibility, night combat is fought at close range, making the deployment/redeployment and maneuver of forces more difficult. Land-based aircraft are one of the most effective combat arms for striking enemy shipping and shipping-related coastal facilities. Their high degree of readiness and maneuverability enable aircraft to concentrate their strikes against transports, warships, or aircraft covering enemy ships at sea.

Command and Control

Major naval operations are most often planned, prepared, and conducted by the numbered fleet commanders or component service/functional commanders (joint/combined force maritime component commanders—JFMCCs/CFMCCs, in U.S. terms) subordinate to a theater commander (commanders). Sometimes operational-tactical commanders (COMJTFs, in U.S. terms) and their staffs can plan and conduct major joint naval operations. Unlike a tactical commander, the operational commander also has the overall responsibility for logistical support and sustainment. Also in contrast to tactical commanders, operational commanders can sequence and synchronize the actions of all the services and elements under their commands.

The decision for a major naval operation requires a so-called long-range estimate. In preparing the estimate the operational commander must take into account all aspects of the situation in a given maritime theater of operations—military, political, economic, sociological, ethnic, and informational—several weeks or even months in advance. Because of the much wider range of uncertainty involved and the larger scope of a major naval operation, a long-range estimate of the situation will necessarily contain a larger number of assumptions. The more assumptions a plan contains, the more likely it will require alteration, modification, or even radical changes (or even to be completely abandoned) during the execution phase.

In contrast to naval battles or engagements, major naval/joint operations are invariably planned. They are normally prepared only when forces have at least local superiority over their opponent, while tactical actions can be conducted regardless of the ratio of forces in a given maritime theater. Major naval/joint operations and campaigns are planned using a "regressive" method, in which the ultimate operational objective is determined first and then major tactical objectives and some minor ones are identified.

Elements

A major naval operation in a high-intensity conflict, regardless of its purpose, comprises three main phases: deployment, combat employment, and postcombat. Depending on the initial bases of operations, one's naval forces may conduct only operational deployment (within a given maritime theater of operations) or, in some cases, strategic deployment (movement from one maritime theater to another) (see figure 2). If forces are deployed forward, as are U.S. Navy forces currently in the western Pacific and in the Persian (Arabian) Gulf, in the case of hostilities they would need to conduct an operational or even just tactical deployment.

The combat phase of a major naval operation would consist of one or more phases, differing in duration. Normally, the area in which combat would take place is divided into a sector of main effort and one or more of secondary effort. The major part of one's naval forces, arbitrarily called "main forces," would be employed in the sector of main effort, while the rest—"supporting forces"—would operate in the secondary sectors. The first phase encompasses the concentration of forces taking part in the operation in the assigned ocean/sea area, followed by a series of maneuvers to obtain a favorable position with respect to the enemy's forces. The main forces would destroy or neutralize the enemy's operational center of gravity, through a series of related major and minor tactical actions. Supporting forces would tie down enemy forces to facilitate the success of the main forces. Normally deception would be conducted in one of the sectors of secondary effort. The cumulative effect of the actions of the main and supporting forces should lead to the accomplishment of the operational objective. Afterward, this operational success must be consolidated by pursuing and destroying the remaining enemy forces. In a major offensive naval operation, the shift from the combat phase to the pursuit phase should be seamless. An operational pause might be required after the end of pursuit to regenerate the combat potential of friendly forces that took part in the operation. After operational success is consolidated, forces might withdraw to

FIGURE 2
Major Naval Operation and Its Main Elements

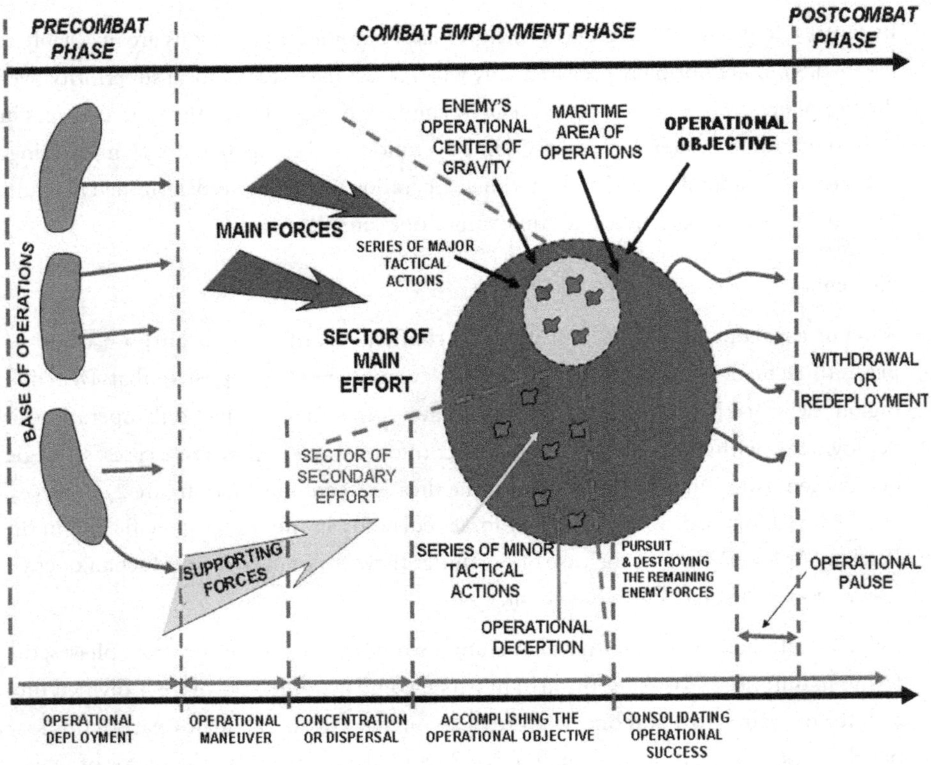

Note: This figure depicts elements of a major naval operation in generic terms only.

their base of operations or be directed to redeploy to carry out other tasks in the same or an adjacent theater.

Prerequisites

The success of a major naval operation is greatly dependent on the proper synchronization of the selected elements of theaterwide support. Sound theater command organization is one of the most important prerequisites for the success of a campaign as a whole and of its subordinate major operations. Other prerequisites for the success of a major naval operation include operational intelligence, operational command-and-control warfare, operational fires, operational logistics, and operational protection.

A major naval operation is the principal method of accomplishing operational objectives in a maritime theater. It can be the most effective way to achieve decisive results within a given time frame, bringing about a drastic change in the situation in a specific theater of operations. The only other way of achieving the same results is attrition warfare, which should be avoided even when the operational commander enjoys quantitative superiority.

Major naval operations are, however, an area of study of operational art that most Western naval theoreticians and planners have generally neglected. Too much emphasis is given instead to the tactical employment of one's naval forces and to various noncombat missions in operations short of war. The real danger is of creating the impression that major naval operations will not be conducted in the future; this could not be more wrong. The absence of a serious threat at sea today should not delude anyone into believing that naval forces will never be required to plan and conduct major operations in the future. In addition, any regional conflict may require the U.S. Navy to plan and execute major naval operations. Threats to national interests at sea tend to come without much warning. Fleets are built and maintained primarily not to conduct operations short of war but to wage war, whether regional or global.

The current highly unsatisfactory situation can continue without adverse consequences for the U.S. Navy and other Western navies only so long as they do not face peer competitors who achieve superiority not through the number of platforms and advanced weapons but by thinking and acting operationally instead of tactically.

Notes

1. *Department of Defense Dictionary of Military and Associated Terms,* Joint Publication [JP] 1-02 (12 April 2001, as amended through 14 April 2006), p. 365. The U.S. Army's Field Manual *Operations,* [FM] 3-0, includes a more detailed explanation concerning what constitutes a major operation, as do *Doctrine for Joint Operations,* JP 3-0 (10 September 2001, and 2005), and *Joint Operation Planning,* JP 5-0 (2006). These documents correctly define a major operation as "a series of tactical actions (battles, engagements, and strikes) conducted by various combat forces of a single or several services coordinated in time and place, to accomplish operational and sometimes strategic objectives in an operational area"; JP 3-0, pp. GL-12 and GL-13; FM 3-0, p. 2-3; and JP 5-0, p. GL-14.
2. Adolf Feil, ed., *Grundlagen der Theorie der operativen kunst der SSK,* trans. Dieter Foerster and Bernd Kulbe (Dresden: Friedrich Engels Military Academy, March 1989; translation of *Operativnoye iskusstvo voyenno-morskovo flota*), p. 24.
3. Hans Tietze, "Zu einigen Kategorien der Theorie der Taktik der Seestreitkraefte," *Militaerwesen* 3 (March 1983), p. 49.
4. Stephen W. Roskill, *The War at Sea 1939–1945,* vol. 2, *The Period of Balance* (London: H.M. Stationery Off., 1956), p. 36.
5. Tietze, "Zu einigen Kategorien der Theorie der Taktik der Seestreitkraefte," p. 48.
6. JP 1-02, in contrast, defines "raid" as "an operation, usually small scale, involving a swift penetration of hostile territory to secure information, confuse the enemy, or to destroy installations; it ends with a planned withdrawal upon completion of the assigned mission" (p. 313).
7. Tietze, "Zu einigen Kategorien der Theorie der Taktik der Seestreitkraefte," p. 48.

8. H. Engelmann, "Die Sicherstellung von Seeoperationen," *Militaerwesen* 3 (March 1980), p. 69; S. Filonov, "Morskaya Operatsiya" [Naval Operation], *Morskoy Sbornik* 10 (October 1977), p. 24.

9. Hermann Franke, ed., *Handbuch der neuzeitlichen Wehrwissenschaften*, vol. 1, *Wehrpolitik und Kriegfuehrung* (1936), (Berlin/Leipzig: Walter de Gruyter, 1936–39), p. 206; Wilhelm Stanger, *Grundzuege der Lehre von der Strategie. Studienbehelf fuer did K.K. Kriegsschule*, vol. 1, *Theorie mit Kuerzeren Bespielen* (Vienna: Verlag der K.K. Kriegsschule, 1884), p. 79.

10. Alfred T. Mahan, *Naval Strategy: Compared and Contrasted with the Principles and Practice of Military Operations on Land* (Boston: Little, Brown, 1911), pp. 31–32.

11. Franke, ed., *Handbuch der neuzeitlichen Wehrwissenschaften*, vol. 1, p. 206.

12. Stanger, *Grundzuege der Lehre von der Strategie*, vol. 1, p. 79.

13. Joseph Rodriguez, Jr., *How to Maximize the Advantages of Interior Lines at the Operational Level* (Fort Leavenworth, Kans.: School of Advanced Military Studies, U.S. Army Command and General Staff College, 1987), p. 5.

14. Including eleven battleships, four large and four light aircraft carriers, four seaplane tenders, eighteen heavy and eight light cruisers, fifty-five destroyers, twenty-three transports with 6,500 landing troops embarked, thirteen tankers, twenty submarines, several smaller vessels, and about 250 carrier-based aircraft.

15. Including three fleet carriers with about 250 embarked aircraft, four heavy cruisers, fifteen destroyers, two oilers, and nineteen submarines; Samuel E. Morison, *History of United States Naval Operations in World War II*, vol. 4, *Coral Sea, Midway, and Submarine Actions, May 1942–August 1942* (Boston: Little, Brown, repr. 1984), p. 76.

16. Juergen Rohwer and Gerhard Huemmelchen, *Chronology of the War at Sea 1939–1945*, 2nd ed. (Annapolis, Md.: Naval Institute Press, 1992), p. 63.

17. Burkhard von Muellenheim-Rechberg, *Battleship* Bismarck: *A Survivor's Story* (Annapolis, Md.: Naval Institute Press, 1980), pp. 264–67.

18. Tomislav Bolfek, "Pomorske Operacije—oblik borbenih dejstava nase Ratne mornarice," *Mornaricki Glasnik* (Belgrade) 5 (September–October 1977), pp. 765–66.

19. Giuseppe Fioravanzo, *A History of Naval Tactical Thought* (Annapolis, Md.: Naval Institute Press, 1970), p. 203.

CHAPTER TWO

Types of Major Naval Operations

In generic terms, the main purposes of a major naval operation today in the case of a high-intensity conflict at sea can be fleet versus fleet (destroy the enemy fleet at sea or in its bases); fleet versus shore (conduct an amphibious landing on the opposed shore and destroy enemy coastal installations/facilities); attack against an enemy's maritime trade; defense and protection of friendly maritime trade; destruction of an enemy's (or protection of friendly) sea-based strategic nuclear forces; and support of ground forces operating in the littoral.

Major Operations: Fleet versus Fleet

Destruction or neutralization of the main enemy forces at sea and in their bases is the main prerequisite for obtaining and then maintaining sea control in a given part of a maritime theater. This is especially important at the start of the hostilities at sea. In the past, this objective was usually accomplished by a clash of major parts of the opposing fleets in a so-called decisive battle or "general fleet action." However, a decisive battle became obsolete in the aftermath of the battle of Jutland in June 1916. During World War II, the Royal Navy, the German Kriegsmarine, the Imperial Japanese Navy (IJN), and the U.S. Navy conducted several major operations aimed to destroy the opposing fleet at sea or in bases. In that conflict most fleet-on-fleet encounters took place when one fleet was providing distant cover and support to a major convoy or amphibious task force, or when the stronger fleet threatened an amphibious landing to lure an inferior force into a decisive battle.

Major naval operations to destroy a neutralize a major part of the enemy fleet can be conducted in a distant ocean area or in a narrow sea. A small but skillfully led and highly trained navy can relatively quickly obtain local control of the sea even against a stronger combination of opponents by conducting a major naval operation, as the Israeli navy did during the nineteen-day Yom Kippur/Ramadan War of 1973. The Israelis secured early control of those parts of the eastern Mediterranean declared by Syria and Egypt as war zones.[1] The Israelis seized the initiative at the outset and inflicted heavy losses on their enemies. In a naval battle off Latakia on the night of 6–7 October, an Israeli force composed of five *Saar*-class missile craft sunk three Syrian Osa missile craft, a torpedo craft, and a minesweeper.[2] In a naval battle between six Israeli *Saars* and Egyptian Osas off Damietta-Baltim on the night of 8–9 October, the Israelis sank three Egyptian missile craft; another was heavily damaged and subsequently destroyed by artillery fire.[3] These two battles radically changed the operational situation at sea, to the Israeli advantage.[4]

In the Gulf War of 1990–91, the U.S. Navy and its coalition partners achieved sea control of the northern part of the Persian (Arabian) Gulf by a series of related strikes and attacks against the Iraqi naval forces between 22 and 24 January 1991. The coalition aircraft destroyed two minelayers, one oiler (acting as scouting ship), two patrol craft, and one hovercraft. On 29 January, in an engagement off Bubiyan Island, U.S. and British missile-armed helicopters and ground-attack aircraft destroyed four patrol craft and ran aground fourteen more that were probably carrying commandos to take part in the Iraqi attack on Kafji; in a separate incident, a British helicopter destroyed a large patrol craft.[5] A day later, the American and British helicopters and ground-attack aircraft attacked a force consisting of one former Kuwaiti patrol craft and three Iraqi amphibious craft and one minesweeper; all suffered damage, in various degrees. In another encounter, eight combat craft, including Osa missile craft, were attacked by U.S. ground-attack aircraft in the northern part of the Gulf; four craft were sunk and three damaged.[6]

In the initial phase of a war in a typical narrow sea, the principal objective of a major naval operation would be to obtain sea control, thereby creating favorable conditions for other operational tasks by the navy and other services. The key prerequisite for success in such a major operation is air superiority in a given part of the maritime theater.[7] The objective of the operation would be accomplished by destroying the enemy's surface forces and submarines; suppressing tactical air forces at enemy airfields; destroying enemy antisubmarine warfare (ASW) forces at sea and in their bases or airfields; suppressing or destroying enemy air defenses; and suppressing or interfering with the work of command posts and electronic surveillance. In the second phase of the operation, fleet forces would strike enemy forces at sea, in their bases, and at airfields.[8]

In a typical narrow sea, it is possible to achieve surprise and thereby complete a major naval operation against the enemy fleet in a very short time. Shallow water allows for the extensive use of mines in both the offense and defense. Because of the short distances involved, forces can be deployed covertly and rather quickly. Short-range land-based aircraft can also be employed.[9]

The stronger side can plan and execute a major operation aimed at destroying or neutralizing a major part of the enemy's fleet at its bases prior to or shortly after the outbreak of the hostilities at sea. In general, four basic methods of destroying an enemy fleet in its bases have been used: strikes from beyond the range of enemy base defenses; penetration of the bases by surface forces; raids by special force teams; and blockade of the sea approaches to enemy naval bases besieged as well by troops on land.[10]

In the past, the best results have been achieved by striking from beyond the effective range of the enemy's defenses. Experience shows that it is not necessary to have numerical superiority for the success of such a major naval operation. In fact, the attacker penetrating the enemy's basing area or delivering strikes from long range will be most likely successful, if he achieved a surprise. Such an attack can be conducted at the opening of the hostilities, when neither side possesses sea control, or in case of disputed control. Strikes against enemy ships in their basing areas and airfields can be carried out simultaneously with destroying enemy ships at sea, as the Japanese did in their surprise attack on Pearl Harbor in December 1941.

An attack on the enemy fleet at its base can also be conducted with only local and temporary sea control in the objective area, as, for example, the Royal Navy did in its successful carrier attack on the Italian main naval base at Tarent (Taranto) on 11–12 November 1940 (Operation JUDGMENT). The British force of one aircraft carrier *(Illustrious),* with twenty-one aircraft, and escorts launched an attack from about 180 miles. The attackers achieved complete surprise, and despite strong Italian defenses heavily damaged one new battleship *(Littorio)* and two older battleships (of the *Giulio Cesare* class), one of which sank, while slightly damaging one heavy cruiser and a destroyer. All but two British aircraft returned safely to their carrier.[11] This attack—drastically, if temporarily—changed the naval situation in the central part of the Mediterranean, to the British advantage.

Land-based aircraft can be highly successful in attacks on an enemy fleet in its base. For example, in 1941 the Luftwaffe's First Air Fleet (Luftflotte 1), responsible for cooperation with Army Group North, assigned a squadron of Stuka dive-bombers specially adapted for attacking ships to make a series of strikes on 16–19, 21, and 23 September. The force sank two Soviet destroyers and one submarine; heavily damaged two battleships (one subsequently sank) and two heavy cruisers; and inflicted light damage on two destroyers, a submarine tender, a submarine, and torpedo craft.[12] The Luftwaffe launched another major effort (Operation EISSTOSS) to eliminate the threat of the remnants of the Soviet Baltic Fleet on 4–5 April 1942.[13] These attacks resulted in heavy damage to one battleship, four cruisers, and one destroyer; a minelaying cruiser, a destroyer, and a training ship suffered light damage.[14] In another series of attacks

(Operation GOETZ VON BERLICHINGEN), carried out in the last week of April, German bombers attacked Soviet ships and shore facilities but further damaged only one cruiser. In the end, despite a substantial effort, the Germans failed in their stated objective to destroy the Soviet surface fleet bottled up in the Gulf of Finland.[15] The Soviet warships, based in the Leningrad–Kronshtadt area, continued to provide fire support to the defenders on the Leningrad front and the Oranienbaum beachhead.[16]

In a more recent example, on 17 January 1991 the U.S. carrier-based aircraft and attack submarines deployed in the Persian (Arabian) Gulf and the Red Sea repeatedly struck Iraqi naval installations near Umm Qasr and Basra with smart bombs and Tomahawk missiles.[17] American and coalition aircraft attacked, between 25 and 28 January, Iraqi ships based in Umm Qasr, the Bubiyan Channel, and the port of Kuwait, sinking one minelayer, two patrol craft, and one transport.[18] In another action, on 4 February, coalition aircraft attacked the Iraqi naval base at Al Kalia and disabled two missile craft. Helicopters from a U.S. frigate engaged four Iraqi patrol craft off Maradin Island, sinking one of them and damaging another. With this action, the threat of the Iraqi navy to the U.S. and coalition ships operating in the northern part of the Gulf practically ended.[19]

Special forces have shown great potential for attacking enemy ships in their bases in narrow seas. Small groups of commandos can be covertly deployed by submarines, fast attack craft, or aircraft in the base's vicinity. In World War II, Italian naval commando teams conducted several successful raids against naval bases, at La Valletta (Malta), Souda Bay (Crete), and Alexandria. The most successful was the attack on Alexandria, on 19 December 1941; three Italian combat swimmers penetrated the harbor and planted limpet mines that eventually sank two British battleships *(Queen Elizabeth* and *Valiant)* and a tanker. This action led to a dramatic, although temporary, change of naval situation in the Mediterranean. The British were left with only three light cruisers in the eastern part of the Mediterranean, and because of the entrance of Japan into the war, no additional battleships or aircraft carriers were available. Fortunately for the British, because the battleships did not capsize, two months passed before the Italians realized what their swimmers had accomplished.[20]

During the Yom Kippur/Ramadan War, Israeli frogmen penetrated Port Said, a major Egyptian port. In the Gulf of Suez, patrol craft penetrated two Egyptian anchorages and entered the main port of Adabia. In the northern part of the Red Sea, Israeli naval commandos penetrated the port of Hurghada and destroyed, without loss to themselves, four Egyptian *Komar*-class missile craft in the Red Sea. These attacks forced the Egyptians to remove their remaining missile craft from the base, which strengthened the Israeli navy's control of the Gulf of Suez. Repeated raids on Hurghada slowed down the flow of oil from the Suez wells to Egypt.[21]

Today, nuclear-powered attack submarines (SSNs), modern conventionally powered submarines, land- and carrier-based aircraft, and surface ships armed with long-range cruise missiles are the most effective platforms for destroying enemy ships in their bases. Attack submarines armed with antiship or land-attack cruise missiles can strike enemy naval bases effectively from distances of several hundred miles. Air strikes against enemy naval bases in an enclosed sea, such as the Persian (Arabian) Gulf, can be far more effective than strikes mounted from the open ocean, because of the much shorter distances and the larger number of land-based aircraft that can be used. These strikes can be conducted at high intensity and repeated at short intervals. In some instances, not only fixed-wing aircraft but also missile-armed helicopters can be effectively employed. Antiship cruise missiles (ASCMs) are today the principal weapons for striking enemy ships in their bases. Torpedoes, bombs, and mines also could be used.

In the near future, it is unlikely that major naval operations will be conducted on the open ocean; a more likely scenario is conflict between the navies of riparian states in an enclosed or semi-enclosed sea, or between a blue-water navy and one operating in the littorals. Hence, attacks on naval bases or anchorages, combined with attacks at sea, seem to be the most likely method open to a stronger navy to destroy or neutralize a smaller fleet. The most likely scenario is that a blue-water navy would try to destroy a major part of the enemy's fleet in its bases prior to or shortly after the outbreak of the hostilities. Such a major naval operation can be conducted in the course of supporting advances by friendly troops along the coast. A stronger side at sea trying to isolate certain sea/ocean areas will find it necessary to destroy a major part of the enemy fleet. In some cases, the weaker side at sea might decide to engage enemy forces at sea in order to weaken or lift a blockade.

Major Operations: Fleet versus Shore

The stronger side at sea will occasionally conduct major naval operations to accomplish operational objectives on the enemy coast, amphibious landings to destroy or annihilate enemy coastal installations/facilities or important military-economic centers deep in the enemy's territory.

Amphibious landing operations are extremely complex to plan and execute. They are inherently joint in character, regardless of their purpose or the size of forces involved. In general, they can be aimed at seizing an area of enemy-controlled coast that gives access to a military operational objective inland; speeding the advance of friendly ground forces along the coast; eliminating or taking control of a large naval base or port; preventing the enemy from capturing a large base or port; cutting off an enemy army's avenue of escape; and countering enemy evacuation attempts across the sea. A credible amphibious capability may also tie down a sizable enemy force in the defense of a coast or of offshore islands.

Prior to the advent of aircraft, one of the main prerequisites for a successful landing on an opposed shore was at least local and temporary sea control in the amphibious objective area. Yet in some rare cases, amphibious landings succeeded without it. For example, in their war with Russia in 1904–1905, the Japanese went ahead with landings in Korea despite the existence of the Russian squadrons in Port Arthur and Vladivostok.

More recently, it was not possible to conduct successful amphibious landings without not only local sea control but some degree of control in the air. In some cases, control of the air was sufficient to ensure success of the landing despite overwhelming enemy strength at sea.[22] For example, the German invasion of Norway in April 1940 (Operation WESERUEBUNG) was successful due largely to the Luftwaffe's control of the air over Skagerrak and Kattegat and of the sea approaches to Norway. The British landing at San Carlos Bay on East Falkland on 21 May 1982 was conducted although the British had been unable to obtain the desired degree of sea control in the objective area because of intermittent attacks by Argentine land-based aircraft against British carrier task forces.

Sometimes, although initial amphibious landings have succeeded, the lack of sufficient forces to control the adjacent sea surface and subsurface greatly complicated efforts to expand the bridgeheads and accomplish the operational objectives. This has happened even when the attacker possessed local control of the air. For example, the Allies seized the initial lodgment on Guadalcanal in August 1942, but their subsequent inability to mount a major operation on land or to secure the sea approaches to the island led to protracted fighting on land, at sea, and in the air. The Allies generally controlled the air in the lower Solomons, but the sea control was disputed each day between sunset and sunrise. The Japanese, who were more proficient in night fighting and torpedo tactics than the Americans, were able to bring reinforcements and supplies to their embattled troops on the island. This attritional phase lasted for almost eight months before the Japanese gave up their attempts to regain control of Guadalcanal. By then, this protracted struggle had considerably slowed down the Allied operational tempo, because no further major landings up the Solomons chain had been possible until Guadalcanal was made secure.

Attacks by naval forces to destroy or annihilate installations/facilities on the enemy coast are usually (but not always) conducted as part of the exercise of sea control. They can be planned either to tie up or divert a major enemy force from the main sector of effort. They can have a significant, if temporary, psychological effect, raising morale among friendly forces and the population and depressing that of the opponent. The main methods of employment of naval forces are strikes and raids; in some cases, a major naval operation might be a preliminary to an amphibious landing operation or part of a much broader naval blockade. Such operations are often conducted by naval forces and air forces. For example, the main purpose of TF 38's attacks in mid-October 1944 was to destroy Japanese ports, shipyards, and airfields in the Ryukyus and on Formosa and Luzon. Today, longer-range and more capable carrier- and land-based aircraft, armed with cruise missiles and smart bombs, make it possible to attack enemy coastal installations and facilities along a large part of the enemy coast and far into the depth of defenses. Major navies today have a much greater ability than in the past to attack a variety of targets well into the enemy's operational and even strategic depth.

One of the largest bombardments of enemy coastal installations in history was carried out by Britain's Force H, from Gibraltar, against the Italian coast. The force consisted of two battleships, one aircraft carrier, one cruiser, and ten destroyers. The attack took place between 31 January and 4 February 1941, when aircraft from the aircraft carrier *Ark Royal* attacked the dam at Lake Tirso in Sardinia. However, poor weather forced Force H's commander, Admiral James F. Somerville, to retire to Gibraltar without striking the ports of the upper Tyrrhenian Sea. On 6 February Force H again got under way, separating during the night of 8–9 February into two groups: one, centered on *Ark Royal,* with three destroyers, bombarded Leghorn (Livorno), while the other group, comprising the rest of the force, shelled installations in the port of Genoa. Damage on shore was heavy, and one Italian battleship *(Duilio)* at Leghorn suffered no hits. The British lost only one aircraft. With their ponderous organization and the lack of interservice cooperation, the Italians were unable to mount a counterattack from the air on the retiring British ships.[23]

In 1973, after obtaining sea control in the first few days of hostilities, the Israelis carried out the second phase of a plan to exercise sea control. Starting on 10–11 October, missile craft and gunboats attacked oil shipping and port facilities at the ports of Latakia, Bâniâs, and Tartus. Between then and the end of hostilities on 17 October, the Israelis conducted almost uninterrupted nightly attacks against Syrian and Egyptian ports (Damietta, Port Said, Rosetta, and Alexandria). Additionally, they attacked many military installations, notably radar stations and surface-to-air missile (SAM) sites along the Egyptian-controlled coast, especially in the northern part of Sinai. These attacks were predominantly conducted by missile craft, but commandos took part as well.[24]

The enormous advances in technology since the end of World War II allow a blue-water navy to attack and destroy or annihilate diverse military, economic, and other targets hundreds or thousands of miles into enemy territory. This capability, exemplified by the ever-greater effective range and endurance of carrier-based aircraft and land-attack cruise missiles, allows a blue-water navy to exert greater influence on the course of war on land. For example, at the start of the offensive phase of the Gulf War on 17 January 1991, nine U.S. guided-missile cruisers and destroyers in the Red Sea and the Arabian Gulf launched theater land-attack missiles (TLAMs) against selected targets in Iraq.[25] TLAM-C/D missiles were used against well defended targets that called for precision warheads. The U.S. Navy also extensively used its two battleships deployed in the Arabian Gulf to strike a range of Iraqi targets on the coast. The battleship *Missouri* fired sixteen-inch shells against prefabricated concrete bunkers along the Kuwaiti coast on 3 February. Three days later another battleship, *Wisconsin,* relieved *Missouri* on the battle line.[26] The battleships fired more than a thousand sixteen-inch shells against the Iraqi positions after the start of the ground offensive on 24 February 1991, forcing the Iraqis to abandon their positions at the coastal port of Ras Al Qualayah.[27]

Major Operations versus Enemy Maritime Trade

Attack on maritime trade in general has always been an important feature of any war at sea. It constitutes a form of pressure on a country dependent on overseas trade for the necessities of life. Attack on maritime trade is usually highly effective against an island nation, but less so against a continental country with well-developed land communications with the neighboring countries.

The main methods of combat employment of naval forces against enemy maritime trade are naval tactical actions and major naval operations. These actions are meant to destroy or neutralize not only the enemy shipping at sea and in ports but also such other elements of maritime trade as shipyards/ship-repair facilities, port installations/facilities, and railroad/road traffic in the littorals.

Major naval operations planned to interrupt or cut off enemy maritime trade will be conducted only when the objective is to destroy or neutralize a major part of the enemy's maritime trade in a given maritime theater. Such an operation can also be conducted in rare cases when there is a need to destroy enemy shipping carrying large troop reinforcements and materiel to an island or major port, or attempting to reinforce or resupply a position or besieged troops. In contrast to major naval operations conducted in the open ocean, those conducted in a narrow sea involve several naval combat arms and combat arms of other services. Surface combatants, submarines, land- or carrier-based aircraft, special forces, and coastal defense forces would take part in such an operation.[28]

Major Naval Operations to Defend/Protect Maritime Trade

Both the stronger and the weaker sides at sea would commit large forces to ensuring an uninterrupted flow of maritime traffic throughout the conflict. Obviously, this operational task is more critical, and also usually more difficult, for the weaker side.

Maritime trade is made secure by organizing the defense not only of commercial shipping at sea but also of all other elements of trade—port terminals, cargo storage depots, shipbuilding and ship-repair facilities, railway/road junctions, and railway/road traffic in the littoral area. In the littorals, this operational task cannot be successful without the closest cooperation between the navy and other services.

For the stronger side at sea, the destruction or neutralization of the enemy forces that pose a threat to maritime trade can best be achieved through a combination of offensive and defensive actions. Optimally, the enemy surface ships and aircraft should be destroyed at their bases or during transits to their operating areas. These objectives can best be accomplished by a series of major offensive naval operations at the very beginning of hostilities. Subsequently, offensive tactical actions and, occasionally, major naval operations are conducted when the operational situation in the theater is favorable.

One of the most effective methods of defending and protecting maritime trade is to plan and execute a major naval operation. Specifically, such an operation is aimed at preventing the interruption or cutoff of shipping carrying troops, weapons and military equipment, and strategic raw materials. In some cases, such an operation can be aimed to protect traffic in coastal waters.[29] Such an operation can be carried out in a single phase, the duration of which depends on the length of the shipping routes, stay of ships at ports and anchorages, and assembly time for convoys. In some cases, smaller convoys and individual ships can sail independently without protection.[30]

In an enclosed sea theater, the small size of the area, short distances, and relatively fixed routes make it much easier for the enemy to detect and track maritime traffic. This also makes it difficult for ships to avoid attack, especially from the air and by fast attack craft. In general, it is more advantageous to have a small number of transports in each convoy but a large number of convoys. Defense and protection of friendly convoys is considerably enhanced by a direct screen, including air cover. The size and composition of the screening force for a convoy can vary considerably. In general, the ratio of escort ships to transports in a convoy should be between 1 and 1.5:2. In some cases, the defense of convoys can be strengthened by operational cover forces at some distance. Distant cover can be provided by deploying submarines and surface combatants in threatened sectors and maintaining land-based aircraft in a high state of readiness. In littorals, coastal missile/gun batteries can be used as distant cover for convoys.

Destruction/Protection of Seaborne Nuclear Deterrent Forces

Today only large navies, specifically those of the United States and Russia, have the capability to mount major operations aimed at destroying the enemy's sea-based strategic nuclear forces—that is, against ballistic-missile submarines (SSBNs) and their supporting elements. Such operations might be focused on destruction of the enemy SSBNs, either at their basing areas, during their transits to or from them, or in patrol zones. Destroying or protecting SSBNs in the conventional phase of war would be highly complex, in terms of timeliness of preparation, the establishment of a maritime theater, and operational buildup; the difficulty of, on one hand, searching for and pursuing SSBNs and, on the other, reinforcing ASW forces and vectoring them to enemy SSBNs; the large size of the sea areas in which enemy forces are located; and the necessity to overcome strong and deeply echeloned ASW and antiair defenses. All this would make difficult the operations of friendly forces.[31] The likelihood of a strong enemy reaction will also complicate reconnaissance and combat support.[32] Likewise, a major naval operation might be conducted to protect sea-based strategic nuclear forces. Such major operations would most likely be carried out predominantly by naval forces—specifically, SSNs in the open ocean and maritime patrol aircraft overhead. However, in areas closer to friendly or the enemy's coast, such operations would be more joint, because air forces would be expected to take part.

Major Operations in Support of Ground Forces on the Coast

One's naval forces can cooperate with ground forces operating along the coast in both the offense and defense. In the broadest terms, naval forces support troops on the coast by providing cover, support, and supply. *Cover* means preventing enemy air, missile, or gunnery strikes, or amphibious landings on the flank or in the rear of friendly ground troops. These tasks are accomplished by destroying forces that threaten friendly ground troops from the sea, as well as enemy amphibious forces at their embarkation areas, in transit, and in their landing areas. *Support* by naval forces encompasses a range of tasks, from destroying important targets on the coast and in the depth of the enemy defenses to attacking maritime traffic in coastal waters flanking troops on the coast. *Supply* includes transport of troops and materiel; seizure of crossings over water obstacles (straits/narrows, river estuaries, etc.) for friendly troops; and evacuation of troops from beaches, naval bases, or ports.

Naval forces can support ground forces by major and minor tactical actions. Only in some situations it will be necessary to plan and conduct major operations, whether predominantly naval or joint, with other services. The success of the ground troops along a coast is affected not only by the naval forces directly supporting them but also by those conducting major operations to destroy enemy naval forces at sea or in their bases and thereby obtaining sea control. In general, a major naval operation can be planned to destroy a fleet threatening friendly troops operating along the coast; to seize, by an amphibious landing, a major island, strait, or a stretch of enemy-held coast; to defend a coast against an enemy amphibious landing; to attack the enemy's maritime trade or protect one's own; and to conduct large-scale evacuations of troops or civilians.[33]

The principal tasks of naval forces in supporting friendly ground forces on the offensive are attacks on the enemy troop concentration and installations/facilities in the depth of enemy's defenses; amphibious landings; destruction of enemy naval forces posing a threat to the advance of friendly troops along the coast; prevention of tactical amphibious landings or raids, and the arrival of enemy reinforcements by sea or land; transport of friendly troops and materiel; capture of naval bases/ports; and support in crossing water obstacles.

Naval forces can significantly help ground forces advancing on the coast by blockading large stretches of an enemy-held coast to prevent reinforcement of the enemy's army. Another method of preventing the arrival of enemy reinforcements to frontline troops is to interdict rail and road communications with strikes by carrier- or land-based air or surface combatants. In general, interdiction is more successful the fewer rail or road lines the enemy army depends upon for bringing up supplies, and the flatter the terrain. For example, the UN interdiction effort in Korea was only partially successful, because the communists used terrain features to their advantage; in fact, they increased the flow of supplies to their frontline troops, despite heavy punishment by UN aircraft. They also had effectively unlimited human resources with which to keep roads and rail lines in operation. Another reason for the incomplete success of UN interdiction was a self-imposed limitation against attacking the actual sources of enemy supplies. In addition, UN aircraft were unable to find and destroy at night or in adverse weather small, individual targets not destroyed in daylight raids.[34]

In World War II, a large number of troops were transported across straits and narrows. For example, during the Normandy invasion about five million Allied troops crossed the English Channel. In the same conflict, the Germans transported about a million and a half troops across the hundred-mile-wide Sicilian Narrows (or Strait of Sicily) and eight hundred thousand across the Danish Straits. During Allied Operation BAYTOWN on 3 September 1943, two divisions of the British XIII Corps of the Eighth Army, carried on twenty-two LSTs and 270 smaller landing craft, landed across the two–ten-mile-wide Strait of Messina on the Calabrian coast near Reggio and Villa San Giovanni.[35] Also, large contingents of troops were redeployed from one theater to another through the Strait of Gibraltar, the Suez Canal, the Strait of Malacca, and the Formosa Strait.

In support of a major defensive operation on the coast, one's naval forces would conduct a series of such tactical actions as providing surface fires, laying defensive mine fields, protecting transport of troops and materiel, conducting amphibious raids, taking part in anti-amphibious defense, and helping defend major naval bases/ports. In some situations, naval forces might also conduct major operations to evacuate friendly troops and civilians.

Naval forces would play the principal role in major anti-amphibious operations. Success in such a major operation would depend on timely detection of the enemy movements prior to the landing, activation of the entire coastal defense system, and timely deployment of forces. A major anti-amphibious operation would normally consist of four main phases: strikes against the assembly areas of the enemy invasion forces; actions to destroy enemy forces at sea; a battle for the bridgehead; and a battle in the depth of the landing objective area. For each phase, the defender should determine an objective to be accomplished.[36]

Slowing an enemy's advance along the coast is closely related to protracted defense of naval bases and major ports. Prolonged defense of a naval base could tie up significant numbers of enemy troops and thereby gain enough time for the front line to be stabilized by building up defenses and bringing in new reinforcements. Sometimes a stubborn defense of even a small naval base can win valuable time for preparing defenses of more important naval bases or ports.

The main tasks of naval forces in defense of naval bases and major ports include providing fire support to friendly troops on the ground, taking part in defense against attacks from the sea and in the air, providing support to tactical landings or raids on the flanks of enemy ground forces, keeping open access to naval bases or ports besieged from the land side, transporting reinforcements of troops and materiel, and evacuating troops and civilians.

During World War II, the large naval base at Sevastopol in the Crimea was defended by the Soviets for about 290 days (29 October 1941–7 July 1942). Soviet surface ships and submarines kept the garrison supplied until the Germans, with complete command of the air, began their final attack on the fortress on 7 June 1942. Seven German and two Romanian divisions took part in the operation, supported by the German VIII Air Corps and 208 batteries of all calibers brought to bear on the Soviet fortifications. The Soviet Black Sea Fleet continued to carry reinforcements and supplies to the fortress, support its defenders with gunfire, and evacuate the wounded.[37]

Naval forces should be prepared to carry out evacuation of friendly troops and civilians when their lines of retreat are cut off on land or isolated in a pocket and threatened with destruction. The problem in this type of support is that the timing is critical, when the naval forces involved may have been given other important tasks.

The British successfully evacuated some forty-three thousand Allied troops (out of a total of fifty-seven thousand men) from Greece in Operation DEMON, between 24 and 29 April 1941. About twenty-seven thousand men were landed on Crete, partly to reinforce it and partly to economize on shipping by using the shortest passage until the evacuation was complete. Another sixteen thousand men were taken directly to Egypt under cover of the British fleet. The evacuation of Allied troops from Greece was exceedingly hazardous, because the Luftwaffe rendered the main embarkation port of Piraeus almost unusable. The Germans enjoyed superiority in the air, which by the time of evacuation had become air supremacy. Nonetheless, the entire effort was carried out in five nights. Because Piraeus was put out of action, the embarkation took place from the beaches, three in Attica and three in the Peloponnesus. There was no air cover for Allied troops except for a handful of long-range fighters operating from Crete. British losses were four transports and two destroyers sunk.[38]

One of the largest and most successful evacuations of troops across a small stretch of water took place in the English Channel in 1940. Despite almost constant attacks by the Luftwaffe and advancing German troops, the British extricated from Dunkirk in Operation DYNAMO about 338,000 troops (including 139,000 French troops). About 240,000 were evacuated from the port of Dunkirk, the rest from nearby beaches. The evacuation took place between 27 May and 4 June 1940.[39]

The Germans conducted several large-scale evacuations of their own troops and civilians in the Black Sea in 1943–44. Despite huge Soviet superiority on the ground, at sea, and in the air, the Germans were highly successful in evacuating their personnel and large quantities of weapons and equipment. Preparations to evacuate the Crimea started on 27 October 1943.[40] On 8 April 1944, when the Soviet offensive to retake the Crimea began, about 130,000 German and seventy thousand Romanian troops were on the peninsula. Another ten thousand were cut off in the Yalta hills but were evacuated by barges. The final Soviet attack on the Crimea started on 7 May 1944, and two days later Hitler finally gave the order to evacuate the peninsula. By 11 May only about twenty thousand German troops remained; two days later the last German ship left the Chersonese Peninsula.[41] In one month about 130,000 Romanian and German soldiers had been evacuated by sea. German losses on the Crimea had been seventy-eight thousand men killed or captured. A total of 190 German warships and merchant vessels took part, sufficient to evacuate 123,000 men one way, though large distances were involved—some two hundred nautical miles separated Sevastopol from Constanţa, the main debarkation port. The Axis losses were surprisingly low.[42] The German navy performed extraordinarily well in the evacuation effort, despite the fact that the Soviets controlled both the sea and the air around the peninsula. Despite their numerical superiority on the ground, in the air, and at sea, the Soviets were unable to hinder the embarkation of troops at Sevastopol.[43]

During the Korean War, UN forces carried out one of the most successful evacuations of troops and civilians in history. On 8 December 1950, with the failure of the UN offensive in the north, and in the face of the Chinese advance toward Wonsan, General Douglas MacArthur ordered the evacuation of the X Army Corps from Hungnam on the eastern coast of Korea.[44] A U.S. 3rd Division task force and a Marine shore party totaling about 3,800 troops, some 1,150 vehicles, some ten thousand tons of cargo, and seven thousand refugees were evacuated from the port of Wonsan on 3 December.[45] Two days later, troops of the South Korean 3rd Division and some 4,800 refugees were evacuated by U.S. ships of Task Force 90 from the port of Songjin. The evacuation from the Hungnam area started on 11 December and officially ended on 26 December. The X Corps and the South Korean I Corps, totaling 105,000 troops, were evacuated in orderly fashion, plus some ninety-one thousand civilians, and sealifted to Pusan and Ulsan (some thirty miles north of Pusan). In addition, 17,500 vehicles and 350,000 tons of cargo were transported by sea on board 192 ships.[46] All serviceable equipment and fuel were taken away. The U.S. Navy provided air support with seven aircraft carriers and gunfire support from thirteen ships. Neither Chinese nor North Korean forces made any serious attempt to interfere with the evacuation of the UN troops.[47]

Conclusion

The types of major naval operations are distinguished by their principal purposes. In the broadest terms, they can be grouped as operations of fleet versus fleet, fleet versus shore, and shore versus fleet. In World War II, major naval operations, amphibious landing operations in particular, involved in some cases hundreds of ships, craft, and aircraft. In the Korean War and afterward, only a relatively few major naval operations were conducted. Because of the steady reduction in the size of the blue-water navies and the tremendous increase in their combat power, major naval operations were conducted with relatively small numbers of ships and aircraft. The likelihood of conducting a major naval operation against the enemy fleet at sea is remote today. The most likely scenario is a major naval operation conducted to destroy or neutralize a major part of the enemy fleet at its bases and at sea. In an enclosed or semi-enclosed sea, such as the Persian (Arabian) Gulf, operations will also include participation of land-based air and in some cases also ground forces. Amphibious landings with operational objectives will be conducted in the future, but they will be much smaller in size than in the past. The blue-water navies, such as the U.S. Navy, will occasionally conduct major operations versus enemy shores to destroy or neutralize important installations/facilities in the littorals and even in the enemy's strategic depth. Major operations to attack enemy maritime trade or to protect friendly maritime trade will be conducted mostly by naval forces and other services jointly employed. Most actions in support of friendly ground forces operating in the coastal area will be tactical in scale; major naval operations will be conducted only in exceptional situations.

Notes

1. Syria declared part of the Mediterranean south of 33° N and east of 34° E as a war zone, while Egypt's war zone encompassed the sea area south of 33° N and east of 29.5° E; Walter Jablonsky, "Die Seekriegfuehrung im vierten Nahostkrieg," *Marine Rundschau* 11 (November 1974), p. 653.
2. Ibid., p. 654.
3. Benyamin Telem, "Die israelischen FK-Schnellboote im Yom-Kippur-Krieg," *Marine Rundschau* 10 (October 1978), p. 640.
4. Between 6 and 22 October, Israeli missile craft sank seven Syrian and five Egyptian missile craft, one Syrian torpedo craft and minesweeper, and two Egyptian patrol craft; Jablonsky, "Die Seekriegfuehrung im vierten Nahostkrieg," pp. 654, 662–63. In the Red Sea, the Egyptian navy lost two missile craft, three patrol craft, and fourteen armed fishing craft, while four combat craft (two torpedo craft and two light patrol craft) were captured in Port of Adabia.
5. Hartmut Zehrer, ed., *Der Golfkonflikt. Dokumentation, Analyse und Bewertung aus militaerischer Sicht* (Herford/Bonn: Verlag E. S. Mittler and Sohn, 1992), p. 197.
6. Ibid., pp. 197–98.
7. Yugoslav Navy, *Upotreba Strategijskih Grupacija* (Privremeneo Uputstvo) (Belgrade: n.d.), p. 97.
8. Gerhard Regner, ed., *Operative Kunst der Seestreitkraefte,* Teil II/1, *Seeoperationen und gemeinsame Handlungen der Flottenkraefte und Truppen der Front in Kuestenrichtung* (Dresden: Friedrich Engels Military Academy, 1981; translation of *Operativnoye iskusstvo voyenno-morskogo flota* [Leningrad: A. S. Grechko Naval Academy]), pp. 4–5.
9. Ibid., p. 5.
10. A. Kalinin and G. Morozov, "Evolutsiya Sposobov Unichtozheniya Sil Flota v Punktakh Bazirovaniya," *Morskoy Sbornik* 6 (June 1988), p. 25.
11. Stephen W. Roskill, *The War at Sea,* vol. 1, *The Defensive* (London: H.M. Stationery Off., 1954), pp. 300–301; Juergen Rohwer and Gerhard Huemmelchen, *Chronology of the War at Sea 1939–1945,* 2nd ed. (Annapolis, Md.: Naval Institute Press, 1992), p. 41.
12. Juergen Rohwer, "Der Minenkrieg im Finnischen Meerbusen, Part II: September—November 1941," *Marine Rundschau* 2 (February 1967), pp. 97–100.
13. Gerhard Huemmelchen "Unternehmen 'Eisstoss'—Der Angriff der Luftflotte 1 gegen die russischen Ostseeflotte im April 1942," *Marine Rundschau* 4 (April 1959), pp. 226, 229.
14. The plan of attack envisaged the following priority of targets: battleships, *Kirov*-class cruisers, the half-completed former German heavy cruiser *Luetzow,* and minelaying cruisers; Rohwer and Huemmelchen, *Chronology of the War at Sea 1939–1945,* p. 134.
15. Between 4 and 30 April 1942, about six hundred aircraft, including 162 Stukas, attacked a variety of ships in the Leningrad–Kronshtadt area; twenty-nine were lost. Huemmelchen "Unternehmen 'Eisstoss,'" pp. 231–32.
16. Friedrich Ruge, *The Soviets as Naval Opponents 1941–1945* (Annapolis, Md.: Naval Institute Press, 1979), p. 24.
17. Edward J. Marolda and Robert J. Schneller, *Shield and Sword: The United States Navy and the Persian Gulf War* (Washington, D.C.: Naval Historical Center, 1998), pp. 181–82.

18. Zehrer, ed., *Der Golfkonflikt,* p. 197.

19. Ibid., p. 198.

20. S. W. C. Pack, *Sea Power in the Mediterranean: A History from the Seventeenth Century to the Present Day* (London: Arthur Barker, 1971), p. 194.

21. Ze'ev Almog, "Israel's Navy Beat the Odds," U.S. Naval Institute *Proceedings,* no. 3 (March 1997), p. 107.

22. Geoffrey Till, ed., *Maritime Strategy and the Nuclear Age,* 2nd ed. (New York: St. Martin's, 1984), p. 117.

23. Raymond de Belot, *The Struggle for the Mediterranean 1939–1945,* trans. James A. Field, Jr. (Princeton, N.J.: Princeton Univ. Press, 1951), pp. 97–98.

24. Jablonsky, "Die Seekriegfuehrung im vierten Nahostkrieg," p. 656.

25. Stanley R. Arthur and Marvin Pokrant, "Desert Storm at Sea," in *Proceedings/Naval Review* (May 1991), p. 83; Michael A. Palmer, "The Navy Did Its Job," U.S. Naval Institute *Proceedings/Naval Review* (May 1991), p. 88; Zehrer, ed., *Der Golfkonflikt,* p. 198.

26. Between 8 and 9 February, *Wisconsin* fired 112 sixteen-inch shells against the Iraqi defenses in southern Kuwait; Palmer, "Navy Did Its Job," pp. 89–90.

27. Arthur and Pokrant, "Desert Storm at Sea," pp. 85–86.

28. Delimir Kolec, "Operacije na jadranskom pomorskom vojistu," *Mornaricki Glasnik* 5 (September–October 1985), pp. 755–56.

29. Ibid., p. 757.

30. Tomislav Bolfek, "Rasprava o podeli i osnovnim karakteristikama operacija na JPV" *Mornaricki Glasnik* 6 (November–December 1978), p. 993.

31. *Operative Kunst der SSK. Seeoperationen* (Dresden: Friedrich Engels Military Academy, 1973; translation of *Operativnoye iskusstvo VMF* [Leningrad: 1970]), pp. 11–12.

32. Ibid., p. 13.

33. *Aktuelle Fragen des Einsatzes der Seestreitkraefte auf geschlossenen Seeschauplaetze* (Dresden: Friedrich Engels Military Academy, Heft 225, 1985), p. 8.

34. Malcolm W. Cagle and Frank A. Mason, *Sea War in Korea* (Annapolis, Md.: Naval Institute Press, 1957), pp. 270–71, 279–80.

35. Juergen Rohwer and Gerhard Huemmelchen, *Chronology of the War at Sea 1939–1945,* vol. 2, *1943–1945* (New York: Arco, 1974), p. 347.

36. Bolfek, "Rasprava o podeli i osnovnim karakteristikama operacija na JPV," p. 986.

37. David Woodward, *The Russians at Sea: A History of the Russian Navy,* pp. 216–17. The Luftwaffe *sank* three Soviet destroyers and a number of smaller warships and transports; Ruge, *The Soviets as Naval Opponents 1941–1945,* p. 79.

38. I. S. O. Playfair et al., *The Mediterranean and Middle East,* vol. II, p. 105; Butler, *Grand Strategy,* ed. J. P. W. Ehrman, vol. 2, *Sept. 1939–June 1941,* p. 457; John Creswell, *Sea Warfare 1941–1945,* rev. ed. (Berkeley: Univ. of California Press, 1967), p. 95; Louis L. Snyder, *The War: A Concise History 1939–1945* (New York: Julian Messner, 1960), p. 157.

39. About twenty-eight thousand troops were evacuated prior to 27 May; some eighty thousand Allied troops went into captivity. Hans Umbreit, "Der Kampf um die Vormachtstellung in Westeuropa," in *Das Deutsche Reich im der Zweite Weltkrieg,* vol. 2, *Die Errichtung der Hegemonie auf dem europaeischen Kontinent* (Stuttgart: Deutsche Verlags Anstalt, 1979), p. 296.

40. The German navy commander in chief, Adm. Karl Doenitz, stated that the Seventeenth Army required at least fifty thousand tons of supplies per month, thirty thousand tons of shipping, and sixty naval ferries. For evacuation of two hundred thousand men, forty days were necessary, twice that many in the case of bad weather; Dietrich von Conrady, "Die Kriegsmarine und der letzte Kampf um Sewastopol im Mai 1944," *Wehrwissenschaftliche Rundschau* 6 (June 1961), p. 315.

41. Between 14 and 26 April about eighty thousand troops and 2,500 tons of materiel were carried from the Crimea to Constanța; by 12 May about twenty-six thousand soldiers and 6,100 wounded had been evacuated, while about 8,100 men had been lost. In addition, some sixteen thousand men, including 6,100 wounded, were transported by air to Romania; Ruge, *The Soviets as Naval Opponents 1941–1945,* p. 130.

42. Some seventy-two German warships and auxiliaries, one Romanian craft, seven steamers, seven tugs, and twenty-one

miscellaneous craft were sunk, mostly by Soviet air attacks; however, the losses of the German and Romanian personnel were relatively high—six to nine thousand men. Juerg Meister, *Der Seekrieg in den osteuropaeischen Gewaessern 1941–45* (Munich: J. F. Lehmans Verlag, 1958), pp. 280–82; Ruge, *The Soviets as Naval Opponents 1941–1945,* p. 132; Michael Salewski, *Die deutsche Seekriegsleitung 1935–1945,* vol. 2, *1942–1945* (Frankfurt, a.M.: Bernard and Graefe, 1970), p. 398.

43. Conrady, "Die Kriegsmarine and der letzte Kampf um Sewastopol im Mai 1944," pp. 320, 330.

44. Some sources categorize the evacuation from Hungnam as a "redeployment"; Cagle and Mason, *Sea War in Korea,* p. 164.

45. Ibid., p. 184; Billy C. Mossman, *United States Army in the Korean War: Ebb and Flow—November 1950–July 1951* (Washington, D.C.: Center for Military History, 1990), p. 165.

46. These included six APAs, six AKAs, twelve TAPs, seventy-six time-charted ships, eighty-one LSTs, and eleven LSDs; Cagle and Mason, *Sea War in Korea,* p. 191.

47. In addition, air force and Marine aircraft lifted some 2,600 men, two hundred vehicles, and 1,300 tons of cargo, and several hundred Korean refugees from nearby Yonpo airfield; ibid., pp. 186, 188; Mossman, *United States Army in the Korean War,* pp. 172–75.

CHAPTER THREE

Decision Making and Planning

The decision-making and planning process for major naval operations is based on the same techniques and procedures as for a major land or other major operation. However, there are some significant differences because of the very nature and characteristics of war at sea as compared to war on land or in the air. The operational decision-making and planning process is complex and time consuming. Yet without detailed planning it is hard to employ one's maritime forces successfully. The more detailed the decision making and planning, the easier it is to modify, alter, or even radically change plans once a maritime campaign or major naval/joint operation is under way.

Making a Decision

One of the principal responsibilities of commanders at any level of command is to make decisions for the combat employment of their forces. Decision making is more an art than a science. The most common method of making a decision is a thorough assessment or estimate of the situation. In military terms, the commander's estimate of the situation is a logical process of reasoning by which a commander considers all the factors affecting a military situation to determine a course of action to accomplish a given mission. The commander has to analyze and compare several friendly courses of action and enemy capabilities, in an attempt to identify the course of action that offers the best chances of accomplishing the mission. The estimate of the situation should be *solely* the responsibility of the commander. His personal involvement is driven by the time available for the estimate, his personal preferences, and the experience and accessibility of his staff.

In general, the larger the scale of the situation, the broader the aspects of a given naval situation that the commander and his staff must evaluate. Hard-to-quantify and unquantifiable elements of the situation play a much more important role in the operational commander's estimate than in the tactical estimate.

The estimate conducted by the operational commander is supplemented by the estimates of the various sections (personnel, intelligence, operations, logistics, civil-military affairs, information operations, C4, etc.). These *staff estimates* (or *supporting estimates*) are subordinate to the estimate conducted by the operational commander, assisted by his staff. Staff estimates provide an assessment of the situation and an analysis of the course of action the commander is considering. Each staff estimate includes a conclusion and a recommended course of action for the respective functional area. Staff estimates are conducted concurrently with the commander's estimate. Selected parts of each supporting estimate should be evaluated by the commander and his staff and then adopted, modified, or rejected. The final estimate of the situation should reflect the commander's best assessment of the operational reality at the time of the plan's execution.

The principal elements of the operational commander's estimate of the situation are the mission, the situation, the commander's intent, enemy capabilities (or courses of action), own or friendly courses of action, and the decision. How these elements are arranged and integrated in the estimate process vary from service to service, and even from commander to commander. There is also no commonly agreed method of conducting the estimate of the situation. In generic terms, the commander's estimate encompasses the following steps: mission analysis, the physical/human environment, the enemy situation and the friendly situation, analysis of the opposing courses of action, comparison of friendly courses of action, and the decision.

The *mission analysis* is the first and most critical part of the estimate of the situation at any level of command. The main results of this step of the estimate are a restated mission, the commander's planning guidance, and a tentative commander's intent. All the subsequent steps of the estimate must be consonant with the commander's restated mission and initial intent. The mission analysis process provides the direction to the commander and the staff, enabling them to focus effectively on the problem at hand. The operational commander is solely responsible for analyzing the mission and stating it in such a way that subordinate commanders can begin their own estimates of the situation, decision making, and planning. He and the planners should also take fully into account the missions of the neighboring friendly commanders during their own mission analysis.

In general, the larger the objective, the more complex the naval situation in terms of space, time, and force. The naval situation is characterized through the statuses and actions of opposing forces, terrain, oceanography, state of weather and time of day and year, and other conditions.[1] It encompasses three overlapping situations: ground, naval, and air (space). The operational commander and his planners should analyze each component of a military situation individually and then collectively create a picture of the situation from a multiservice or multinational perspective.

The estimate of the naval situation encompasses the estimate of the three distinctive but interrelated and overlapping situations: the friendly situation, the enemy situation, and the physical (including human) environment. It might also be necessary to estimate the situation of neutrals. In generic terms, estimating the situation means analyzing the causality of all its elements and their influence on the accomplishment of the mission. For that purpose, the operational commander should divide the situation into its constituent parts and determine how each part affects the employment of enemy and friendly forces, and in some cases neutrals as well. He also must determine how to make maximum use of favorable factors in the situation for the employment of friendly forces.

The analysis of the *physical/human environment* differs significantly at each level of naval command. A naval tactical commander is concerned with those aspects of the physical environment that affect the course and outcome of a naval tactical action, such as a battle or engagement. Hence, he has to focus on those aspects that affect the employment of individual ships, submarines, aircraft, tactical-sized units, and the respective naval combat arm. A naval tactical commander has to assess the physical environment in terms of oceanographic/hydrographic conditions, characteristics of the immediate coastal area, and the weather/climate. He is also concerned with the human aspects of the coastal area. In contrast, a naval operational commander needs to analyze operational features of the environment—that is, those that affect the employment of several naval combat arms and forces of other services taking part in a major naval operation. This task is not easy, because it requires proper synthesis of the tactical information on the physical environment to create an operational picture of the situation. The human factor can either facilitate or restrict the employment of maritime forces in the littoral area. Hence, the possible effects of population, urbanization, ethnicity, religion, and domestic and foreign policy on the employment of friendly and the enemy maritime forces should be evaluated as well.

A proper analysis of the physical/human environment by the naval operational commander includes its advantages and disadvantages for friendly and enemy command and control. A general description of the geography/terrain should be followed by an assessment of its specific aspects and then a conclusion for each. Advantages and disadvantages of the terrain for both friendly and enemy maritime forces should be evaluated in some detail. In the conclusion, the naval operational commander should highlight in which way the geography/terrain and weather/climate is favorable for the intended operation—or conditionally favorable, or unfavorable.

The estimate of the *enemy's situation* is an important part of the naval operational commander's estimate of the situation. Without such an estimate it is simply not possible to assess the enemy commander's true intent and the enemy's capabilities. In general, the naval operational commander and his staff can focus on either the enemy's *capabilities* (ECs) or the enemy's *intentions* (ECOAs, or enemy courses of action). The method of relying on the enemy's capabilities should be used at operational and higher levels of command, whereas reliance on an estimate of the enemy's intentions offers more advantages at the tactical level, where speed and energy in action are much more important.[2] The most important advantage of relying on the enemy's capabilities is that the operational commander and his staff would cover the full spectrum of what the enemy can physically do in a given situation; hence, the risk of being surprised by some overlooked enemy capability is relatively low. Among the disadvantages of relying on the enemy's capabilities is that the enemy commander's intent is not taken into account.

The estimate of the enemy's capabilities is the heart of the estimate of the enemy situation. The U.S. term "enemy capabilities" refers to the actions of which the enemy is physically capable that, if adopted, would affect the accomplishment of one's mission.[3] The term "enemy courses of action" implies that one has reasonably accurate knowledge of what the enemy will do. Enemy intentions are rarely known with any confidence without knowing the enemy's mission—and that information is unlikely to be known to the commander. Even if such information were available, the enemy could change or feign his intentions. No enemy capability should be dismissed or overlooked because it seems unlikely or uncommon. Hence, this term should not be used in the estimate process unless one does in fact have accurate and reliable sources of information on what the enemy's real intentions are; it would be foolhardy to ignore such information. The enemy's intentions are the actions that he might or might not adopt.

The worst thing to do is to determine an enemy capability by "mirror imaging"—thinking that the enemy will not do certain things because one would not do them oneself in a similar situation. The enemy's intentions are subject to change with little or no warning. Hence, the operational commander should focus on the enemy's capabilities, not his intentions. The exception to this, again, is if the enemy's intentions are known and based on accurate and relevant intelligence. For instance, Admiral Chester W. Nimitz, Commander in Chief, Pacific Ocean Area/Commander in Chief, Pacific Fleet (CINCPOA/CINCPac), and his staff obtained through MAGIC intercepts precise and reliable information on what the Japanese plan was prior to the battle of Midway in June 1942. They used that information to concentrate available carrier forces off Midway. The enemy's ability to change doctrinal procedures and techniques based on lessons learned should be considered as well. The operational commander should be interested in everything the enemy can do that can materially influence his course of action.[4]

The main input on the enemy's capabilities should be provided by the intelligence estimate. However, the operational commander and his staff, not the intelligence section/division, should make the final determination of the enemy's capabilities (or courses of action). Because of the nature of their work, intelligence analysts tend to rely more on the enemy's intentions than on his capabilities. This viewpoint is more likely to result in fewer potential enemy courses of action, because many options are—intentionally or unintentionally—omitted from consideration.

Generally, in estimating enemy forces the operational commander should focus on those forces that can be employed against friendly forces. Among other things, this includes information on the enemy's overall strength, combat effectiveness, and combat activity up to the time of the estimate. It should also include time calculations regarding the effect of geography/terrain and weather/climate.[5]

The process of determining the full range of the enemy's capabilities is time consuming. The operational commander has to possess complete information about the enemy and to estimate the enemy situation comprehensively. This method is also more difficult to learn.[6] The method of relying on the enemy's intentions in assessing the enemy's situation is much faster. It favors quick and energetic action. However, not all information on the enemy is collected and evaluated. Thus, the risk of omitting some enemy capability is correspondingly higher than in other methods. This method can also be misused to justify a decision based on the enemy's favorable posture for friendly forces.[7]

Subordinate major tactical commanders in conducting their estimates of the situation should use the enemy's capabilities developed by the naval operational commander as the basis for developing the enemy capabilities pertaining to their respective forces. Then each of the enemy's major tactical elements would become the main or supported force, while other components would act in support. The same logic should be applied in developing courses of action for friendly forces.

In estimating the *friendly situation* the naval operational commander and his staff should have precise knowledge of where the friendly forces are, which of them can be employed immediately and which will become available, and which forces of the adjacent commanders can be counted on or must be supported by other forces.[8] This part of the estimate should be focused on the key elements of the factor of force: order of battle (OOB), combat training and readiness, leadership qualities, state of morale and discipline, and logistic support and sustainment. The operational commander should ask himself what the solutions are for accomplishing his mission, which one solution is possible, and which one appears desirable.[9] He should also evaluate the effect of the factor of time on the employment of friendly forces, and he should draw certain conclusions from the estimate of friendly forces.[10]

In developing each friendly course of action, the operational commander and his staff should envision the employment of friendly forces and resources as *a whole,* taking into account planning assumptions determined in the mission analysis step of the estimate, externally imposed limitations, the military situation, and the previously reached conclusion on the features of the physical environment. He also has to consider the effect of rules of engagement (ROE) on the employment of friendly forces. The emphasis should be—and this is often forgotten—on the *entire* mission.[11] Hence, the operational commander and his staff should make sure that each friendly course of action encompasses the phases within a specific major phase of a major naval operation.

The first task in developing the friendly course of action is to determine whether the mission's purpose (objective) is offensive, defensive, or a combination of the two. Afterward, the operational commander and his staff should identify friendly forces one or two command echelons down. Major force elements should be expressed in generic terms, thereby leaving the subordinate tactical commanders to assign force elements with their proper designations. At the operational level, the focus should be on the force that would accomplish the main objective of the pending operation. Afterward, each tactical commander in conducting his estimate should use friendly courses of action developed by the operational commander as the basis and develop his course of action in more detail. The tactical commanders would also focus on those subordinate force elements that would accomplish the main objective.

Each friendly course of action should be fundamentally different from all others; otherwise, the operational commander and staff will waste time analyzing a larger number of friendly courses of action than warranted by the mission and the situation. As with developing the enemy's capabilities, the *variability,* or *distinctiveness,* of a course of action is ensured by varying the employment of friendly combat forces in terms of the factors of space, time, and forces. Taken together, the courses of action should exhaust the possibilities for meaningful action by the friendly forces. If these criteria are met, the friendly courses of action can be considered collectively exhaustive and mutually exclusive.

A friendly course of action developed by the operational commander should include deployment of a force as a whole, followed by possible actions against enemy forces, by answering the questions of *who* (force element), *what* (type of action), and *where* (location of action). At this step, the estimate process should not include *when* or *how* the particular task should be carried out. The temptation to use too many words in describing a friendly course of action should also be avoided. The more concise the statement of the possible courses of action, the more clearly distinctions among them can be recognized. All the friendly courses of action developed during the estimate process, except the final one, are tentative solutions to the problem at hand.[12]

The *analysis of friendly courses of action and the enemy's capabilities (or courses of action)* is one of the most important parts of the entire estimate process. This step compares, for the first time, friendly courses of action with enemy capabilities (or courses of action). The main purpose of this dynamic interaction is to eliminate friendly courses of action least likely to accomplish the mission. This analysis should include considerations of the weather, oceanography/hydrography, terrain, and friendly and enemy forces; relative strength; composition and dispositions; logistic support; and requirements for future operations. Very often it is necessary for a maritime operational commander also to consider fully the effects of political and other nonmilitary aspects of the situation.

The operational commander and his staff need to analyze and compare both the enemy's capabilities (or courses of action) and friendly courses of action in a dynamic setting. This part of the estimate can be extremely complex, because many intangible elements on both sides fall within the scope of the analysis. The operational commander's judgment and experience are critical in selecting which friendly course of action should be retained and which discarded. The key to this determination is reliable and thorough knowledge of both the enemy's and friendly capabilities in terms of such hard-to-quantify factors as leadership, doctrine, morale and discipline, unit cohesion, and degree of jointness.

Very often, the operational commander and his staff would use quantifiable methods in evaluating combat potential for each friendly course of action and enemy capability (or course of action).[13] At the same time, however, the operational commander should not allow mathematical computations to drive his analysis. Calculations of such factors as the enemy's intentions, will to fight, morale and discipline, etc., are of limited usefulness because of the inherent difficulties in properly assessing and anticipating the results of the highly dynamic interactions between the physical elements of the situation and its human actions/reactions.

In conducting the *comparison of friendly COAs* the operational commander and his staff make a list and consider the advantages and disadvantages of each retained friendly course of action, identify actions to overcome disadvantages, make final tests for feasibility and acceptability, weigh the relative merits of each friendly course of action, and then select one that offers the greatest chances of accomplishing the mission. Because the advantages and disadvantages of any particular friendly course of action can be numerous, one method is to concentrate on factors that "govern" the outcome of the visualized interactions with the enemy. These governing factors are those aspects of the situation or externally imposed considerations that the commander deems decisive to the accomplishment of his mission. The factors can range from selected principles of war (e.g., surprise, mass, offensive, economy of effort, security, etc.) to command and control, logistics, protection, deception, potential losses for friendly forces, a particular oceanographic feature or geographic position, and lines of operations. They might also include factors that facilitate or even ensure the success of future operations. However, the true effectiveness of this method is suspect, because the selection of governing factors and their respective numerical weighting are almost entirely arbitrary.

After identifying advantages and disadvantages for each of the retained friendly courses of action, the operational commander and his staff should identify what actions to take to minimize the disadvantages. These can include a novel method of employing forces, deception, or requests for additional forces or air support. This process is applied to each of the retained friendly courses of action. A final test of adequacy and acceptability should be conducted before the final determination is made as to the optimal and second-best friendly courses of action.

The operational commander relies heavily on professional judgment and experience in making the final selection of a friendly course of action. If none of the friendly courses of action are adequate or acceptable, the operational commander should present them, along with supporting facts, to the next-higher commander. He should point out what can be accomplished under the circumstances and estimate what additional forces would be required to accomplish the original mission. It is then the responsibility of the higher commander either to order that the selected course of action be carried out despite the consequences or to change the original mission statement.

The *operational decision* is the final and most important step in the operational commander's estimate of the situation. It must clearly show the commander's will to win.[14] It should be a clear, logical, concise statement of the operational commander's intentions. His subordinates should never be in doubt about what they have to do. The decision should spell out what the command as a whole must do and provide information as to *who, when, where, how,* and *why.* The operational commander may decide to adopt a course of action without modification, alter the best friendly course of action to overcome disadvantages, develop a new course of action by combining favorable elements of multiple courses of action, or discard all courses of action and repeat the mission analysis and course-of-action development.

The soundness of the decision determines the effectiveness of the resulting action; both are dependent on the professional judgment, fortified by knowledge and founded on experience, of the commander. The operational commander's professional judgment is greatly strengthened by mental exercise in applying a logical process to the solution of military problems.

A written decision should contain three distinctive elements: the decision statement, the commander's intent, and the operational idea (scheme). The *decision statement* should express in broad terms what the force as a whole has to do and why. It should be written in the first person, clearly, concisely, and in commonly accepted and understood doctrinal terms. Before issuing the decision statement, the operational commander and his staff should review the initial *intent* and, if the situation has changed, modify or refine it before issuing the statement. The operational commander's *intent* provides the framework for the initial planning for the staff and subordinate commanders. The intent is one of the most critical tools of successful command and control. It is not simply a restatement of a mission.[15] In its simplest terms, the commander's intent is a concise description of the military conditions (or military "landscape") the commander wants to see *after* the specific mission as a whole is accomplished. At the operational level, the intent is an expression of the commander's operational vision. It provides a link between the mission and the operational idea (scheme). The operational commander should consult with his chief of staff and other members of the staff before issuing his intent to subordinate commanders.[16] It should be issued to subordinate commanders at the *next two levels down*.[17]

The operational commander's intent should inform subordinate commanders of what needs to be done if the initially issued orders become obsolete because of unexpected changes in the situation.[18] The intent should provide insight into why the operational commander is embarking on a particular course of action.[19] The operational commander's intent should define mission success in a way that establishes commonality of purpose and unity of effort. It should allow the subordinates to exercise initiative should the original plan or order no longer apply or unexpected opportunities arise.[20]

The intent clarifies but does not simply restate the purpose element (the why) of the mission. In it the operational commander focuses rather on the situation beyond the immediate phase of a major naval operation. This helps subordinates pursue the mission without further orders even when events do not unfold as planned.[21] The broader the commander's intent, the larger the framework and the greater the room for the subordinate tactical commanders to exercise freedom of action in accomplishing the assigned mission.

The operational commander should not include in his intent the method by which subordinate tactical commanders should accomplish their assigned missions. Such details invariably limit subordinate tactical commanders' ability to act creatively in accomplishing the mission. Moreover, they prevent subordinates from exercising initiative in reacting to sudden and unexpected changes in the situation. Also, the intent should not include an operational idea (scheme) for the pending major naval operation.

The operational commander's intent should be clearly expressed in its entirety in the execution part of a plan/order. His intent should be written in the first-person singular, using compelling language. It should fully reflect the personality of the commander issuing it. The intent should also be complete, telling subordinates what they must do and why. In addition, it should define success, in terms that are executable.[22]

The operational commander's intent should be concise, so the subordinate commander can remember it.[23] Optimally, it should be no longer than four or five sentences. The commander's intent should also be concise, to eliminate verbiage that could lead to ambiguity or misinterpretation. It should define priorities. It should use precise and commonly understood doctrinal terms whenever possible.

The main purpose of developing an *operational idea* is to clarify further the operational commander's restated mission and intent. The operational idea should explain in some detail when, where, how, and why each service or functional component (for a maritime campaign) or combat arm (for a major naval/joint operation) would be employed to accomplish the mission, as well as who and what is involved. It should contain sufficient detail to allow planners to draft operation plans and orders—but not so much as to limit flexibility during the subsequent planning process.

In drafting the tentative operational idea, the commander and his staff should express in broad, clear terms the tasks and objectives to be accomplished by each major force element for each phase of the forthcoming operation. However, the focus should be on the first phase of a major naval operation. The operational idea should also clearly state in geographic terms the sectors of main and secondary effort and the point of main attack or defense. It should clearly indicate the sequencing of tasks and the tentative time frame for each.

After the decision is made, the next problem is the organization of forces carrying out operational deployment and the direct preparation of forces and units for the start of combat actions. These tasks are extremely difficult to resolve effectively in a narrow sea, due to the size of a theater and modern reconnaissance and surveillance. This requires thorough measures of operational cover, concealment, and deception so as to lead the enemy in false directions or sectors.

Planning

In general, *operational planning* deals with preparing, testing, and executing plans for a specific campaign/major operation by using tenets of operational art. Like strategic planning, it is conducted continuously, both in peacetime and in time of conflict or war. Operational planning is necessarily much broader in terms of the factors of space and force, and it looks much farther ahead than tactical planning in terms of the factor of time.

In generic terms, the basic types of planning are sequential, parallel (or concurrent), and collaborative. In *sequential* planning, the higher commander and his staff conduct planning first, followed by the planning of subordinate tactical commanders. This type of planning is usually detailed and methodical. It minimizes the risk that subordinate headquarters will receive an obsolete estimate of the situation and plans. At the same time, the entire planning process takes a relatively long time to complete. In *parallel* planning, the operational commander and the subordinate commander start the planning process almost simultaneously. This type of planning results in great savings of badly needed time. It is highly suitable when the time available is short. Today, parallel planning is facilitated by continuous information sharing between higher and subordinate headquarters. In addition, early, continuous, and rapid sharing of planning information among subordinate and adjacent forces allows subordinate commanders to start planning concurrently with higher headquarters. This is facilitated by issuing a warning order and planning guidance.

In a *collaborative* planning process, there is a real-time interaction among the commands and staffs of two or more echelons that are developing plans for a single maritime campaign or major naval/joint operation. This type of planning greatly speeds up the entire process. It allows subordinates to provide the higher commander with their current assessments and statuses and apprise him of how they are postured for various operations. Collaborative planning is made possible by information technologies that allow the real-time exchange of information by voice and video. It greatly enhances the understanding of the commander's intent and planning guidance throughout the force. It also greatly reduces the time required for all echelons to complete a plan. Collaborative planning is a very dynamic process. Information sent by the higher commander might change often, which could cause confusion for subordinates. Commanders might be inhibited in adjusting plans by a concern that subordinate commanders will think plans are being changed too often. Additionally, there is a great danger of a disconnect between the higher commander's and subordinate commanders' plans.

Planning Responsibilities

Major naval operations should be planned, prepared, and executed by a single commander who has the necessary authority to sequence and synchronize the employment and sustainment of several naval combat arms. The coherence of the plans and orders is adversely affected if the commanders and staffs involved are changed often. In case of a regional or general war, a maritime theater of operations command should be established to accomplish a single strategic objective, through planning and executing a single maritime campaign. Subordinate service or functional component commanders normally plan and execute major naval operations as integral parts of a maritime campaign. For the near future, maritime campaigns are unlikely unless there is a conflict between two strong opponents at sea. Hence, major naval operations will be planned by the operational commanders as part of a land campaign encompassing the littorals.

Planning Considerations

In planning a campaign major naval operation, operational commanders should consider a number of military and nonmilitary aspects of the situation in a given maritime theater. In planning a campaign the operational commander should thoroughly evaluate the effect employment of friendly forces would have on the nonmilitary aspects (political, diplomatic, economic, religious, legal, environmental, informational, and other) of the strategic situation in the theater. While foreign policy or purely domestic politics should not dominate the entire operational planning process, the operational commander and his planners should never neglect or willfully ignore the foreign-policy ramifications of a certain course of action.

A major naval operation cannot be planned without fully considering the legal aspects of the use of force. Therefore, a sound operation plan should conform to the law of armed conflict and, in the case of conflict or war in a maritime theater, the law of the sea. The national political leadership should resolve the issue of how forces can be most effectively employed while observing international law and the law of armed conflict. ROE should be timely and carefully drafted to guide the operational commander and his staff during the planning and execution of a maritime campaign or major naval/joint operation. While forces must act in compliance with the law of armed conflict, they have an inherent right to individual and unit self-defense. ROE must therefore be clear, tailored to the situation, and reviewed for legal sufficiency. At the same time, operational commanders should avoid the other extreme—crafting too-strict rules of engagement that might make battlefield operation extremely difficult for friendly forces or even preclude success.

A plan for a major naval operation should be based on a number of *operational* considerations, arbitrarily also called *operational design*. Plans for a maritime campaign and those for a major naval operation have some commonalities, but they also differ significantly in terms of their ultimate objectives and corresponding factors of space, time, and force. A sound operational design should ensure that forces are employed in a logical and coherent manner and focused on the assigned operational or strategic objectives. Operational planners should contemplate several time- and space-related phases.

They should firmly identify required resources and direct them toward the accomplishment of the assigned ultimate objectives.[24]

The basic plan for a major naval operation contains, in rudimentary form, only the most important elements of an operational design. Other elements of the operational design are detailed in the annexes to the basic operation plan and in the plans of subordinate tactical commanders. In generic terms, the design of a major naval operation should envisage all the actions and measures required for the deployment, combat employment, and redeployment of one's maritime forces.

In operational terms, the heart of the design for a major naval operation is the operational idea (or scheme). During the planning process, the provisionally drafted operational idea in the operational commander's decision should be refined and, if necessary, modified to meet possible changes in the operational situation.

Normally, it is the responsibility of the national or alliance/coalition leadership to decide when a new campaign should start. However, the operational commander and his staff should provide input to the strategic leadership concerning the optimal time, in terms of the day of the year and time of day, to start the campaign. The key issues pertaining to the factor of time that should be satisfactorily resolved include the anticipated duration of a campaign or major naval operation, time for planning and preparation, mobilization time, and time required for deployment of combat forces into the theater.

Major Phases

A major naval operation is normally an integral part of a maritime or land campaign. Regardless of its purpose, it comprises three main phases: deployment, combat employment, and postcombat. *Reconstitution* of maritime forces might be necessary in the aftermath of a campaign or major naval operation in case the losses in ships, aircraft, and personnel are so high that units or force elements can no longer conduct their assigned combat missions. Reconstitution involves two distinctive tasks, *reorganization* and *regeneration of combat power*. Specifically, the operational commander has to restore combat-attrited units under his command to a level of combat effectiveness commensurate with mission requirements and the availability of resources. This includes cross-leveling or replacing personnel, supplies, and equipment; reestablishing command and control; and conducting essential training.

Perhaps the most critical step in designing a major naval operation is to determine how long will be required for the accomplishment and then consolidation of the ultimate operational objective. The tentative length of each major phase of a major naval operation should then be determined.

Relations with Media

The operational commanders must also consider media impact during the planning and execution of a major naval operation. Planning should facilitate the interaction between the military and national and international press organizations. Relations with the media are another important area that requires the urgent and continuous attention of the operational commander and subordinate tactical commanders and their staffs. Operational commanders should integrate into their plans the effects of media coverage of their operations. Relations with media should not be solely the responsibility of the public affairs officer.[25]

The public affairs officer should, however, be actively involved in the planning process. The aim should be to use public affairs to reduce the adverse impact of information flow on the course and outcome of a major naval operation. Today, the instantaneous nature of information dissemination enables both opposing operational commanders to modify decisions and actions very quickly. The information flow cannot be controlled by either side, but it can be managed through careful and synchronized issuance of releases designed to shape the information environment.[26]

Control and Coordination

The operational commander and his staff should also make the necessary arrangements for *control* and *coordination*. These include determination of the boundaries of the surface area, to facilitate coordinating and deconflicting operations. In war at sea, a boundary is a line by which areas assigned to adjacent units and formations are defined. Lateral, rear, and forward boundaries can be used to define areas of operations for land and naval forces. Various fire-coordination lines must also be precisely delineated. Air sorties are not constrained by surface boundaries in a general sense. However, because all forces use airspace, the operational commander should deconflict the various uses of it. In addition, the delivery of air weapons inside surface boundaries normally requires coordination with the surface force commander.

Interagency Cooperation/Coordination

One of the most critical, yet also most difficult, problems to resolve satisfactorily in planning and executing a major naval operation is ensuring smooth and effective *interagency coordination.* This entails comprehensive cooperation between the operational commander and his subordinate tactical commanders and other government and private organizations. Coordination of the plan for a major naval operation with adjacent operational commanders and supporting commanders is an obvious necessity. In addition, the operational commander might need to plan for coordination with various nongovernmental and private volunteer organizations, especially in operations short of war.

Basic Plans and Orders

In generic terms, the planning process ends with the completion of the basic *operation plan* (OPLAN). Because a maritime campaign can last several weeks or even months, sound operation plans should contemplate possible changes in the situation on both sides within a given time frame. The farther ahead one projects, the less certain and detailed a plan should be. Therefore, the plan describes only the initial phase of a major naval operation with some certainty. Normally, for the first phase of a major naval operation, an *operation order* (OPORD) is written. An operation plan is based on a number of planning assumptions developed or modified during the operational commander's estimate-of-the-situation process. The basic operation plan also contains several annexes to deal with specific parts of the plan that require elaboration (e.g., task organization, operations, intelligence, IO/C2W, logistics, personnel, public affairs, civil affairs, C4). An overview of the plan for a major naval operation should be presented graphically on the nautical chart. The key part of the graphic presentation should be a broad outline of the operational ideas for the entire major naval operation.

The main body of the plan should be written clearly and concisely. At the same time, the plan should provide sufficient detail for subordinate component commanders to write their own plans and orders. Short plans are preferable to those that are too detailed. The less directive and detailed the plan, the more freedom of action is allowed to subordinate commanders.

The plan for a major naval operation should reflect the commander's operational vision. It should be focused on the ultimate objective to be accomplished. It should ensure the proper sequencing of objectives or tasks and synchronize the employment of all subordinate forces. It should describe in some detail the overall concept and the operational commander's intent, a tentative sequence of phases that would lead to success, and a general concept for key supporting functions.

A sound plan for a major naval operation should focus on defeating or neutralizing the enemy's operational center of gravity. Yet neither the plan nor an operation order should explicitly spell out what the operational center of gravity is. It is even worse for the operational commander and his staff to determine tactical centers of gravity in their plan. These and similar details are best left to subordinate commanders.

The plan should be as complete as possible; otherwise, its execution will invariably end as a series of improvisations, with significant loss of time and resources. In general, the plan for a major naval operation should be based on a few essential tenets of operational warfare. The simplest things are usually the most successful. A plan should be flexible, in case something goes wrong during its accomplishment. In war, there are always unforeseen circumstances that require changes, often drastic, in existing plans. Hence, operational commanders and their staffs must be able to adjust their plans quickly without benefit of a detailed estimate of the situation or thorough planning.

The timeliness of operational deployment of maritime combat forces into the theater depends upon sound planning prior to the start of a major naval operation. The *deployment plan* is closely related to logistics plans. All these plans must be approximately in harmony.

Perhaps, there is no greater error than having several commanders and their staffs plan deployments, especially when headquarters are physically separated by large distances. Only the operational commander can properly plan and execute deployment for a major naval operation.

The deployment concept developed during the operational commander's estimate of the situation is the basis for deployment planning. The commander's decision lays out in some detail the deployment sequence of subordinate force components. Afterward, subordinate tactical commanders should be given responsibility to sequence the deployment of their force elements on the basis of the operational commander's sequencing timetable. The operational commander and subordinate component commanders should evaluate the time required to complete the deployment, the timing of the deployment, and logistic sustainment requirements.

Combat forces should be deployed in a timely manner, because the course of the entire major naval operation is greatly affected by the initial deployment of forces. Sometimes sudden crises require the redeployment of maritime forces to other parts of the same theater or into a different maritime theater.

The operational commander and his staff should develop a provisional *organizational concept* during the estimate-of-the-situation process. This includes, among other things, the proposed boundaries of the maritime area of operations and its subdivisions, command relationships, and forces' subordination. Optimally, the operational commander delegates sufficient authority and responsibility to subordinate component commanders so they can exercise necessary freedom of action.

During the planning phase, the provisional concept of *task organization* is further refined and included as an annex to the OPLAN. Task organization should not be confused with the operational command structure. In contrast to the latter, task organization is established for a specific mission and situation. It is a product of the operational commander's estimate of the situation. The determination of task organization depends principally on the mission assigned to the force as a whole; the nature and scope of combat and, often, nonmilitary actions; the geographic characteristics of the area; friendly and the enemy's forces; and the time available. Normally, task organization should not undergo major changes in the course of a major operation.

Supporting Plans

One of the most important supporting plans for a major naval operation is the *logistic concept,* developed concurrently and in harmony with the commander's operational idea. Logistic support and sustainment are the most critical prerequisites for the successful deployment of one's maritime forces. The logistics plan is critical for supporting and sustaining maritime forces in all phases of the pending major naval operation. It should provide a rough outline of the logistic requirements during the deployment of one's forces, the major combat phase, and the post–major combat phase.

The plan for logistic support and sustainment should be fully integrated with the basic plan for a major naval operation. The operational commander is primarily responsible for developing the plan to use logistic resources to support the execution of the operational idea. In addition, the operational commander coordinates the logistic part of the plan with the service components, and other commands supporting the major naval operation. The logistics plan should be flexible and make provision for anticipated changes in the situation. Logistics planners should not focus only on the deployment of combat forces but should also make plans for support and sustainment for all anticipated phases of a major naval operation. This helps to avoid major interruptions in logistic support and sustainment to friendly forces.

The *deception plan* is an integral part of the plan for a major naval operation; it requires early and detailed attention. Deception planning should result in a plausible, realistic plan within the limits of the resources available. It should outline the methods selected for conveying the deception story to the target and ensure that all means are considered. Deception plans are highly classified. Their existence and distribution must be kept secret; otherwise, the entire deception effort might end in failure and even undermine the success of the basic plan. Sometimes a deception plan is drafted in the course of a major naval operation.

The basic plan for a major naval operation should never depend upon the success of a deception plan. However, a sound deception plan should help the execution and ultimate success of the basic plan considerably. A deception plan entails a certain degree of risk to the success of the basic plan. Hence, the operational commander and his staff should properly evaluate the degree of that risk during the planning process. Deception planners should consider how the planned deception effort fits not only the operational but the political and diplomatic schemes as well.[27]

A plan for a major naval operation sometimes requires a separate scheme for conducting *operational fires.* The main purpose of operational fires is to establish favorable terms for the initial major naval operations in a maritime campaign. Properly planned, operational fires can have a devastating effect on enemy forces. These fires are one of the best means available to the operational commander to define and shape the battlefield. A failure of operational fires will usually require readjustments or even alterations in the plan for a major naval operation. The operation plan should provide for proper sequencing and synchronization of operational and tactical fires.

A *plan for operational intelligence* should be also prepared as an integral part of planning for a major naval operation. Without solid intelligence support, it is difficult to develop sound plans for deployment, combat employment, and redeployment. Intelligence plays also an especially critical role in the preparation of operational deception plans. Intelligence support should be continuous during the entire planning process. Plans for a major naval operation should contemplate timely and effective intelligence preparation of the maritime area of operations and effective dissemination of intelligence throughout the chain of command. In a major naval operation, the rapid and systematic exchange of intelligence information is essential. A particularly difficult task is collecting data on the enemy's intentions and possible reaction to actions.

A plan for a major naval operation today should be supported by a *plan for operational IO and C2W.* Properly planned and executed, C2W can deliver a decisive blow against the enemy's information structure at the outset of hostilities. An IO/C2W plan has an essential role in the targeting process for the entire force. Electronic warfare and psychological operations consume much time and resources, so preparations for IO/C2W should begin well before the start of the planned major naval operation.[28]

Planning offensive C2W depends on the assigned mission, the commander's intent and operational idea, and synchronization with the plan for a major naval operation. The aim is to seize and maintain the initiative by degrading the enemy's informational systems and forcing the enemy to react. Offensive C2W requires sound balancing and integration of all its components. The planners must also consider the effect that degradation of the enemy command and control may have on ongoing efforts to influence the enemy operational commander's actions. Sometimes degrading the enemy C4 links too severely adversely affects these efforts.[29]

To attack the enemy C4 systems, planners must understand how the nodes within that system fit together to make the enemy system work. Within a C4 system, one or more nodes may be of such critical importance that their destruction immediately degrades the functioning of the entire system. To provide a viable target, the C4 node should be not only critical but also vulnerable—that is, it should be susceptible to degradation or have an exploitable weakness. An enemy C4 node must be accessible—there must be some way to reach it. Finally, degrading it should be feasible.[30]

The operational commander and his staff must also synchronize selected elements of *operational protection* to enhance the chances of the successful conduct and outcome of a major naval operation. Because resources are finite, the operational commander should find a balance between the forces and assets required for force protection and, especially, those necessary for the protection of the friendly operational center of gravity. He also must try to do everything possible to protect adequately selected elements of the logistic infrastructure in the theater, C4 nodes, information systems, naval and air bases, shipyards and ship-repair facilities, and various civilian installations and facilities. The list of what should be protected is a long one. This is especially the case in a developed theater. Yet the operational commander's ability to provide adequate protection is usually very limited. Efforts to protect everything invariably end in protecting nothing. Hence, it is an art to determine what needs to be protected and to what extent and then to distribute the available forces and assets judiciously to do the job. Operational commanders must integrate operations security (OPSEC) measures into the planning process to protect the major naval operation from the enemy's attempts to obtain information on its details and the time of its execution. This requires a series of security measures conducted over time throughout the chain of command. Effective OPSEC minimizes the signature of the joint force and helps to avoid set patterns. When these patterns cannot be changed, various cover and concealment measures should be used to deceive the enemy.

Synchronization of Operational Functions

Operational commanders and their staffs should also make necessary and timely arrangements for *synchronization of operational functions* in support of a maritime campaign. Because of the much smaller scale of the ultimate objective, a major naval operation requires synchronization primarily of combat support and some parts of each operational function. The operational command organization is the key element that can either facilitate or derail synchronization of all other operational functions. Hence, the higher political-military leadership must ensure that the operational command organization is timely and properly established. Security and timeliness of information is also influenced by successful functioning of naval communications. However, the limited dimensions of a narrow sea provide especially favorable conditions for effective electronic warfare. This can seriously threaten the security of not only communications but the entire information display system.

Conclusion

The estimate of the situation is the bedrock on which the operational commander's decisions are made. The format of the estimate is not as important as the process of reaching a sound operational decision. The estimate does not guarantee a sound decision, but it does ensure that all elements of the situation are addressed and certain conclusions drawn. It is in the estimate process that the operational commander must show his ability to think broadly, rather than about tactical details. The operational commander depends on the input of many staff officers and subordinates to obtain the best possible knowledge of the situation. This is why he must insist on such qualities as personal integrity and good judgment in his chief of staff and subordinate commanders.

The estimate of the situation should follow a logical and commonsense path and should ensure that no principal element of the situation with a bearing on the decision is omitted. The key prerequisites for a sound decision are the operational commander's knowledge, judgment, and understanding of all the elements of the estimate of the situation.

The process leading to a sound decision is largely an art rather than a science. However, planning for a maritime campaign or major naval/joint operation is both an art and a science. Many elements of planning—for instance, combat forces' capabilities, techniques and procedures, deployment planning, and logistic planning—can be expressed in quantifiable terms. The art part of planning encompasses the operational commander's knowledge and understanding of the complexity of the dynamic relationship between friendly and enemy forces. It also includes balancing ends, means, and ways throughout the planning process. It involves determining the ultimate and intermediate objectives, allocating resources among competing and often contradictory requirements, identifying the enemy's and the friendly centers of gravity, selecting forms of maneuver, conducting deception, establishing a sound task organization, and addressing many other elements of sound plans. The operational commander's willingness to accept high but prudent risk is another part of the "art" aspect of the planning process.

Planning for a maritime campaign or major naval/joint operation should result in a plan with built-in flexibility. The initial clash with enemy forces invariably forces the operational commander to adjust, modify, or completely rewrite his plan. The enemy cannot be made to conform to one's will, at least not initially. This fact, combined with the inherent problems of execution of a maritime campaign or major naval/joint operation, results in a continuous fog of war and friction until the assigned operational or strategic objectives are accomplished. Hence, the operational planning process does not end with writing a plan and issuing orders to subordinate commanders.

In the course of the execution of a maritime campaign or major naval/joint operation, the operational commander should continuously monitor the situation and adjust his plans and orders accordingly. Frequent changes of plans should be avoided. More important, the principle of unity of command should be fully observed. Simplicity requires one to avoid the complex, complicated, and intricate; the plan should proceed in the most direct and natural manner for the attainment of the assigned operational or strategic objective. Other factors being equal, the simplest plan executed promptly is preferable to a complex plan executed later.

Notes

1. *Methodik der Entschlussfassung durch den Fliegerkommandeur* (Dresden: Militaer-akademie Friedrich Engels, November 1971; translation of *Metodika prinyatiya resheniya aviatsonnym kommandirom* [Moscow: Ministry of Defense, 1970]), p. 21.
2. MaximilianThomas, "Modelle zur Lagebeurteilung. Ihre Entwicklung und Weiterentwicklung," *Truppenpraxis,* 1 (January 1982), p. 8.
3. Jack D. Nicholas, George B. Pickett, and William O. Spears, *The Joint and Combined Staff Officer's Manual* (Harrisburg, Pa.: Stackpole, 1st ed., 1959), pp. 132–33.
4. Department of Operations, *Sound Military Decision* (Newport, R.I.: Naval War College, 1942), p. 140.
5. Middeldorf, *Fuehrung und Gefecht. Grundriss der Taktik* (Franfurt, a.M.: Bernard and Graefe Verlag fuer Wehrwesen, 1968), p. 64.
6. Ibid.
7. Thomas, "Modelle zur Lagebeurteilung," p. 8.
8. Middeldorf, *Fuehrung und Gefecht,* p. 57.
9. Helmuth Greiner and Joachim Degener, *Taktik im Rahmen des verstaerkten Infanterie-Batallions* (Berlin: Verlag "Offene Worte," 1937), p. 26.
10. Rainer Oestmann, *Dazu Befehle Ich . . . ! Handbuch fuer Militaerische Fuehrer* (Regensburg/Bonn: Walhalla, 1998), p. 2402.
11. Nicholas, Pickett, and Spears, *The Joint and Combined Staff Officer's Manual,* p. 110.
12. U.S. Naval War College, *Sound Military Decision,* p. 88.
13. Norman M. Wade, *The Battle Staff,* 2nd rev. ed. (Savannah, Ga.: Lightning, 2005), pp. 2–28.
14. Konrad von Beuningen, "Vom Auftrag zum Entschluss," *Truppenpraxis* 7 (July 1959), p. 525.
15. Glenn, "The Commander's Intent: Keep It Short," p. 51.
16. Walter N. Anderson, "Commander's Intent: Theory and Practice," *Armor* (May–June 1998), p. 47.
17. Glenn, "The Commander's Intent," pp. 52–53.
18. Lawrence G. Shattuck, "Communicating Intent and Imparting Presence," *Military Review* (March–April 2000), p. 67; Edward J. Filiberti, "Command, Control and the Commander's Intent," *Military Review* 8 (August 1987), p. 56.
19. Gary P. Petrole, *Understanding the Operational Effect* (Fort Leavenworth, Kans.: School of Advanced Military Studies, U.S. Army Command and General Staff College, 8 May 1991), p. 4; Anderson, "Commander's Intent," p. 47.
20. Anderson, "Commander's Intent," pp. 46–47.
21. Ibid., p. 47.
22. Ibid.
23. Filiberti, "Command, Control and the Commander's Intent," p. 55; Glenn, "The Commander's Intent," pp. 52–53.
24. Fuehrungs akademie der Bundeswehr, *Operative Leitlinie fuer die Fuehrung von Landstreitkraeften* (Hamburg: Fuehrungs-akademie der Bundeswehr, August 1992).

25. Derik W. Crotts, "Operational Implications of Public Affairs: Factors, Functions, and Challenges of the Information Battlefield," *IOsphere* (Winter 2006), p. 15.

26. Ibid., p. 13.

27. Thomas A. Savoia, *Deception at the Operational Level of War* (Fort Leavenworth, Kans.: School of Advanced Military Studies, U.S. Army Command and General Staff College, May 1986), p. 5.

28. H. Engelmann, "Die sicherstellung von Seeoperationen," *Militaerwesen* 3 (March 1980), p. 71.

29. Elizabeth A. Hurst, *Shaping the Battlefield with Command and Control Warfare* (Fort Leavenworth, Kans.: U.S. Army Command and General Staff College, 7 June 1996), pp. 22–23.

30. Ibid., pp. 20–21.

CHAPTER FOUR

Operational Design

The basic plan for a major naval/joint operation contains, in rudimentary form, only the most important elements of an operational design. Other elements of the operational design are provided in detail in the annexes to the basic operation plan and the plans of subordinate component/functional commanders. The *operational design* is a loose collection of diverse elements that the operational commander and his staff should consider in developing the basic plan. In generic terms, the design for a major naval/joint operation may contain the following: ultimate and intermediate objectives, force requirements, a balancing of operational factors against the objectives, identification of the enemy and friendly operational centers of gravity, initial lines of operations, direction (axis), the operational idea, and operational sustainment (see figure 3).

FIGURE 3
Design for a Major Naval Operation

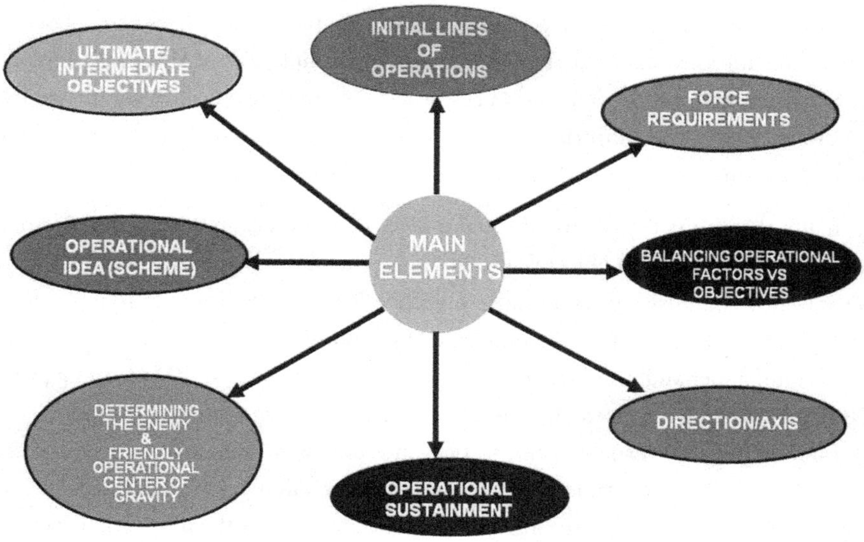

The Ultimate/Intermediate Objectives

The first and the most important step in designing a major naval operation is to determine and articulate properly its *ultimate* and *intermediate* objectives. In general, determination of the achievable military objective seems a simple and straightforward process. However, a number of factors must be correctly evaluated by the naval operational commander and the planners in order to identify militarily achievable objectives. Among other things, the naval operational commander must properly assess the enemy's factors of space, time, and force from the enemy's perspective. In general, the larger the scale of the objective to be accomplished, the larger these factors are and the more difficult it is to evaluate them properly. The naval operational commander needs to evaluate the physical size and operational features of a given sea/ocean and adjacent land areas, the distance from the given physical objective to the staging/deployment area of friendly naval forces, and the time required for the enemy's forces to be deployed, concentrated, and reinforced. The process of determining the objective in a littoral area would also require considering certain nonmilitary aspects of the situation.

Because a major naval operation is an integral part of a maritime or land campaign, its ultimate objective is normally operational and physical in its content. For example, the operational objective of the Japanese attack on Pearl Harbor on 7 December 1941 was to destroy a major part of the Pacific Fleet based on Oahu, aircraft on the ground, and associated naval facilities. The ultimate objective of the amphibious landing at Inchon on 15 September 1950 was to cut off logistical supply lines to the North Korean troops besieging the United Nations forces at the Pusan perimeter and thereby force them to retreat and fight on a reverse front.

Sometimes the ultimate objective of a major naval operation comprises two operational and additional major tactical objectives. For example, in the Allied invasion of the Marianas in June 1944 (Operation FORAGER) the ultimate objective comprised two operational objectives (capturing the Saipan Tinian, and Aguijan island group and the island of Guam) plus one major tactical objective (the island of Rota).

Normally, the ultimate objective of a major naval operation cannot be achieved with a single action. Hence, each objective should be divided into several tasks that should collectively accomplish a given ultimate objective. These, in turn, are translated into the corresponding objectives to be accomplished by subordinate commanders. The operational commander and his planners use a "regressive" (or "backward") planning process in determining intermediate objectives. In designing a major naval operation, several intermediate—usually major tactical and some minor tactical objectives—have to be determined. For example, the Allied amphibious landing on Guadalcanal on 7 August 1942 encompassed three intermediate objectives; all of them were major tactical objectives in scale. Specifically, the Allies had to initially seize the beachhead and the adjacent airfield (later renamed Henderson Field), obtain control of airspace over Guadalcanal and adjacent positions, and obtain control of the sea around Guadalcanal and the adjacent sea area. For each of these objectives there were several minor objectives. For the amphibious forces, minor tactical objectives were the capture of Tulagi and Ndeni in the Santa Cruz Islands. In the Allied amphibious landing at Anzio in January 1944 (Operation SHINGLE), the capture of the towns of Anzio and Nettuno represented two separate but related major tactical objectives, while seizing control of the three smaller towns of Campoleone, Carroceto, and Aprilia constituted a minor tactical objective. In the course of the operation, holding the seven-mile-deep and fifteen-mile-long lodgment was an operational objective for the Allies.

In selecting intermediate objectives, naval operational commanders and their staffs should avoid being either too rigid or too predictable; both would make it easier for the enemy to take timely counteraction. Normally, intermediate objectives should not be pursued simultaneously or in too rapid a sequence; if they are, a major naval operation would be unnecessarily prolonged or the entire effort may fail. Too many intermediate objectives would slow down the tempo of a major naval operation and waste sorely needed time and resources. On the other hand, too few objectives would risk reaching the point of culmination before the specific intermediate objective is accomplished. Planners must find a way to avoid both extremes.

The number of intermediate objectives and the sequence in which they should be accomplished should not be rigidly laid out. Among other things, the degree of the enemy resistance or new intelligence on the enemy force might require change in the number or scale of intermediate objectives; some of them might be replaced with new ones or might simply be bypassed. Also, some intermediate objectives may require neutralization instead of destruction or capture. The number and scale of the intermediate objectives affect considerably the forces' size and mix, the time required for accomplishing a given ultimate objective, and hence the tempo of a major naval operation.

Force Requirements

An important element of operational design is calculation of the overall size and force mix for the entire campaign or major naval/joint operation. The principal factors in this process are the type of the operation, combat potentials of friendly and enemy forces, the number and scale of intermediate objectives and their sequencing, the distances between the base of operations and the prospective operating area, and weather and climatological conditions. In addition, intelligence and logistics play a significant role in determining the size and composition of forces in a major naval operation. This process is an art rather than a science (although quantifiable methods can play a significant role). The naval operational commander's judgment and experience are often the decisive factor in making the final determination on the size and composition of forces to take part in a major naval operation.

In general, the more capable the forces, the fewer ships/submarines or aircraft or troops will be required to accomplish a given objective. The asymmetrical employment of multiservice forces would generally allow smaller forces, as would good intelligence, deception, and surprise. The timeline for a major naval operation can be considerably shortened if the intermediate objectives are accomplished simultaneously or nearly simultaneously. However, in that case, forces must be significantly larger than if the same objectives are accomplished in sequence.

The Allied amphibious landing at Anzio was essentially a failure because the ultimate objective of the operation was in a serious disconnect with the forces assigned to its execution. Only two divisions, the U.S. 3rd and the British 1st infantry divisions, plus the U.S. 504th and 509th parachute infantry regiments, the 2nd British Special Service Brigade with two commando battalions, and the U.S. 6615th Ranger Force (Provisional), totaling some twenty-four thousand men, were assigned to the initial assault. This force would be eventually increased to 110,000 by additional divisions and associated units from the British Eighth Army.[1] However, the key for success was sufficient combat strength at the beginning of the operation, not after the German defenders were able to move significant forces to the lodgment area. Additionally, there were barely sufficient landing craft for a two-divisional lift, plus supporting troops. Only seven days of supplies would be disembarked on the beaches before the tank landing ships (LSTs) had to be withdrawn to England for the pending invasion of Normandy. Another major problem was uncertainty as to whether the weak landing forces would be able to resist German attacks; the Fifth Army was pinned down on the main front and did not have much chance of linking with the Allied forces at Anzio for the first ten days after their landing.[2]

Balancing Operational Factors against the Objectives

The balancing of operational factors versus objective in a major naval/joint operation is governed by the framework of the maritime or land campaign. Any serious disconnect or mismatch between the ultimate or intermediate objective and the corresponding time-space-force factors could greatly complicate and even endanger the success of the entire operation. If the imbalance cannot be resolved, then the objective should be changed or scaled down and brought into harmony with operational factors. This process is complicated and time consuming—again, more an art than a science. In practice, operational factors will rarely be completely, or even approximately, in harmony with one another or with the assigned ultimate or intermediate objective.

In harmonizing friendly operational factors against the respective objective, all considerations should start with quantifiable factors—that is, space and time.[3] In general, the factor of time is the more dynamic and changeable. Over the years, however, the factor of time has been gradually compressed, while that of space has steadily expanded.[4] Normally, space and time can be calculated with a high degree of confidence. The factor of force, however, is often difficult to evaluate properly, because of the presence of many elements that are hard or impossible to quantify. A significant change in any of these factors will invariably disturb the overall balance and require a reassessment of all factors.

In the course of a major naval operation, the operational commander should periodically balance the operational factors against his objective. The higher the intensity of combat, the more often the situation will change, which in turn will require rebalancing operational factors with the objective. Also, any change in objective will require quick reevaluation of the pertinent operational factors and their balancing with the newly determined objective. Naval operational commanders should also fully take into account the effect of information on the factors of space, time, and force, individually and then collectively.[5]

The factors of space and force in a maritime theater can be balanced by reducing the number or scale of the objectives to be accomplished, by, for example, limiting efforts to obtain sea control to a much smaller sea/ocean area or to a single or two physical mediums (e.g., surface and air but not subsurface, or surface and subsurface but not air), increasing the size or combat potential of one's naval forces, or reducing the number of ships/aircraft deployed in a secondary theater of effort. Among other things, the factors of space and time can be brought into balance by operating from shorter lines of operations, employing highly mobile forces, deploying naval forces closer to the scene of potential conflict, prepositioning weapons/equipment and logistical supplies, reducing the size of the operating area, restricting oneself to obtaining temporary instead of permanent sea control in a given sea/ocean area, achieving operational surprise, conducting operational or strategic deception, or reducing the forces assigned for combat support.

Identification of the Enemy/Friendly Operational Centers of Gravity

After determining the ultimate objective of a major naval/joint operation, the operational commander and his planners must determine the corresponding enemy and friendly operational centers of gravity. Both the enemy and the friendly tactical centers of gravity should be identified for each pending intermediate objective. The determination of the proper centers of gravity is one of the most important factors in terms of clarity of purpose and thereby economy of effort in planning and accomplishing the assigned ultimate or intermediate objective of a major naval operation.

A sound plan for a major naval operation should be clearly focused on destroying or neutralizing the enemy's operational center of gravity. At the same time, the friendly center of gravity must be adequately protected; otherwise, the accomplishment of the respective objective will require far more time and resources than are available—or the entire expedition may even end in failure. Therefore, a great deal of thinking and discussion between the naval operational commander and the planners should take place before the enemy and friendly centers of gravity can be considered properly identified.

In naval warfare, a center of gravity can range from a direct screen of a convoy, a surface strike group, carrier group, or amphibious task force, to a major part of the surface forces in numbered or theater fleets. Likewise, in naval air warfare, a center of gravity can be the element of a force of combat aircraft that has the highest combat potential/power, such as a fighter or attack or bomber squadron, or the entire force of fighters and ground-based air defenses, or bombers deployed in a given part of a maritime theater.

In netted forces, computer networks might also represent a major part of the enemy or friendly operational center of gravity. Computer networks are fundamentally different from traditional centers of gravity, in part because they lack the attributes of firepower and maneuver. At the same time, however, a computer network is potentially highly vulnerable to disabling attack.

In some cases, an operational center of gravity may be located beyond the boundaries of the theater of operations in which a new maritime or land campaign is to begin. For instance, the Japanese plan for the invasion of the Philippines in December 1941 contemplated the destruction or neutralization of the major part of the U.S. Pacific Fleet—the operational center of gravity for the Japanese Combined Fleet—based in Pearl Harbor, Hawaii, prior to the attack on the Philippines itself. The Japanese objective was to cripple a major part of the U.S. Pacific battle fleet and thereby prevent it from interfering with the landings in the Philippines. The Japanese attacked Pearl Harbor in the early hours of 7 December, and U.S. air and naval bases in the Philippines one day later.

In a major naval operation, for each intermediate (operational) objective to be accomplished there is a corresponding center of gravity. These centers of gravity are separated in terms of space and time. They have to be attacked, simultaneously and/or sequentially, if a given operational center of gravity is to be ultimately defeated or neutralized. In a campaign conducted in two or three physical mediums (land, sea, airspace), not all operational objectives are equally important; hence, significant differences exist in the relative importance of the corresponding centers of gravity. The objectives at sea and in the air support operational objectives on land. They are in fact the prerequisites for accomplishing the principal operational objective.

The protracted nature of maritime trade warfare (attack on and defense/protection of maritime trade) is largely a consequence of the fact that each side in a conflict avoids massing forces at the operational level. Therefore, in an attack on the enemy's maritime trade or in defense and protection of friendly maritime trade, the principal methods of combat employment of naval forces are major and minor tactical actions. The operational and ultimately strategic objectives are then accomplished over time. A similar situation exists in conducting theaterwide air defenses. In each case, there are generally only a few opportunities to conduct major naval or air operations.

Sometimes, what constitutes a proper center of gravity at the beginning of the war, or during a major part of it, can change due to the cumulative effect of the actions of friendly forces. Normally, in a fleet-versus-fleet operation the enemy's center of gravity is the part of the fleet with the most combat power. For example, in the battles of the Coral Sea, Midway, and the Philippine Sea, the fast aircraft carrier groups were the real operational centers of gravity for both the Japanese and U.S. sides. However, in the Leyte operation the commander of the U.S. Third Fleet, Admiral William F. Halsey, apparently believed that the most serious threat—in modern terms, the operational center of gravity—to his Task Force 38 (fast carrier forces) was posed by the Japanese Mobile Force, Main Body (fast carrier force, under Vice Admiral Jisaburo Ozawa), not the battleships and heavy cruisers of the First Diversionary Attack Force (Vice Admiral Takeo Kurita). One can contend that Halsey's judgment was clouded by his obsession with fighting a decisive naval battle against enemy aircraft carriers, coupled with deliberately vague orders from his superior, Admiral Chester W. Nimitz, Commander in Chief, Pacific Ocean Area (CINCPOA)/Commander in Chief, Pacific Fleet (CINCPAC). Nevertheless, Halsey did not sufficiently account for the declining performance of Japanese pilots after the battle of Midway. By October 1944, the Japanese carriers did not represent as large a threat to U.S. forces at Leyte as did Kurita's heavy surface force.

The enemy's center of gravity can also shift as the major naval operation enters into a new phase. This is usually the case in the execution of a major amphibious landing operation. During the transit phase at sea, the initial main task is defense and protection of the amphibious task force. Hence, the operational center of gravity in the transit phase is the force in distant cover and support. In most cases, such a force is a carrier group or heavy surface ship task force. Very often, it is argued that the amphibious task force is the proper center of gravity for the defender—that were this force destroyed, the attacker could not conduct the planned amphibious assault. This is only superficially true. However, the attacker would not plan an amphibious landing unless he firmly believes that he has sufficient combat potential to overcome the defender's resistance. Normally, the attacker would deploy his amphibious task force and the force for distant cover and support within mutually supporting distances at all times. Moreover, the amphibious task force at sea cannot threaten an enemy's operational center of gravity on land, but the operational covering force can. Thus, for the defender, the primary objective initially is to destroy or neutralize the attacker's distant cover and support, as the U.S. Navy did in the battles of the Coral Sea and Midway in May and June 1942, respectively. In both cases, the Japanese commanders intuitively knew that the planned amphibious landing could not proceed, because even if the amphibious force successfully landed, no naval force remained available that would be sufficiently strong to defend and sustain the landing force ashore.

In the Falklands/Malvinas conflict of 1982, the proper operational center of gravity for the Argentines, prior to the British landing on the Falklands, comprised the two British carrier forces *(Hermes, Invincible).* Without these carriers and accompanying escorts, the British could not have landed. While the loss on 25 May of the thirteen-thousand-ton containership *Atlantic Conveyor,* with its embarked aircraft (ten Wessex and four giant Chinook helicopters) and all the tentage for the landing force, was a serious blow to the British effort, that ship was not a center of gravity for the Argentines but one of the British critical weaknesses/vulnerabilities.[6] The aircraft and equipment could be replaced relatively quickly, but not the aircraft carriers. After the landing, the British 3 Commando Brigade (despite its relatively small size) was the proper operational center of gravity for the Argentine defenders. For the British, the enemy's operational center of gravity was not the Argentine carrier group but the land-based fighter-bombers armed with Exocet missiles. After the landing, the operational center of gravity for the British forces shifted to the Argentine troops defending Port Stanley.

If the attacker lacks sufficient strength at sea, the operational center of gravity might be represented by the land-based fighter aircraft providing cover to the amphibious forces at sea. For example, in the German invasion of Norway and Denmark in April 1940, the Luftwaffe's fighters, not the German heavy surface forces, were the proper operational center of gravity for the Allies. Once the amphibious force starts landing ashore, the main task shifts to defeating the enemy forces and then securing the initial lodgment ashore. Then the attacker's center of gravity shifts to that part of the landing force that has the highest combat potential, usually armor or mechanized forces.

In some cases, an operational center of gravity can shift from one force to another during the sea phase of the operation, because of the combination of the enemy's strength and geography. This is sometimes the case in a major convoy operation conducted in a typical narrow sea. For example, in the Allied major naval operation to resupply the beleaguered island of Malta in mid-August 1942 (Operation PEDESTAL), for defense of the convoy (thirteen freighters and one tanker) was assigned a substantial force of aircraft carriers and heavy surface ships, destroyers, and auxiliaries.[7] Force Z, consisting of two battleships *(Nelson, Rodney)*, three large aircraft carriers *(Eagle, Indomitable,* and *Victorious)*, three cruisers *(Sirius, Phoebe,* and *Charybdis)*, and twelve destroyers, was deployed in the convoy's rear.[8] Force X, composed of two cruisers *(Nigeria, Manchester)*, one antiaircraft cruiser *(Cairo)*, twelve destroyers, and one tug, was deployed as the convoy's direct screen.[9] The convoy and supporting forces departed Gibraltar on 10 August. In the first leg of the convoy's voyage, from Gibraltar to the Skerki Channel, the Allied operational center of gravity was clearly Force Z.

Because of the high threat—posed by the German Luftwaffe and the Italian aircraft based on Sicily, Sardinia, and in North Africa, and by the expected presence of the enemy submarines and mines—it was too dangerous for Force Z to operate in the hundred-mile-wide Sicilian Narrows. Hence, Force Z reversed its course and headed for Gibraltar in the evening on 13 August.[10] From then on and until the convoy reached the approaches of the narrows, the operational center of gravity shifted to Force X. In the third leg of the convoy's voyage, from the Sicilian Narrows to the port of La Valletta, the minesweeping flotilla based at Malta took over the escort of the convoy.[11] However, the Allied fighters based on Malta then became a new operational center of gravity. Despite these shifts in the operational center of gravity, the ultimate objective of PEDESTAL remained the same: bringing the maximum amount of badly needed supplies and fuel to Malta.

Initial Lines of Operations

The initial geostrategic position determines whether one's naval forces and aircraft will operate along interior and shorter lines of operation or along exterior and longer lines of operations. The operational commander and the planners need to evaluate properly the advantages and disadvantages of each in designing the initial major naval operations. In general, operations on multiple and interior lines favor the employment of smaller naval forces. One of the requirements for successful use of exterior lines is that forces move at a higher speed than the opposing forces. The movement of one's naval forces along exterior lines also requires sound synchronization of movement and actions. This is not easy to do, especially if the distance between a force's base of operations and the physical objective is too great. Operations from an exterior position also require that the operational commander seize the initiative.[12]

Direction (Axis)

An important element of a design for a major offensive naval operation aimed at seizure of a part of the coast, large island, or archipelago is a *direction* or *axis*—a swath of land and sea/ocean area and airspace above it extending from one's base of operations to the ultimate physical objective via selected intermediate objectives.

A major naval operation aimed to capture a part of the enemy's coast, island, or strait can be conducted, depending on the extent of the base of operations for friendly forces, along a single operational or several tactical axes. An operational axis extends from one's base of operations via several intermediate, usually tactically significant, positions to the last intermediate objective in a projected major naval operation. An operational axis consists of several tactical axes.

One of the advantages of using a single operational axis of advance is that it makes it easier to synchronize logistics and operations. By capturing successive island positions, operational reach is extended. This, in turn, allows for consolidation of tactical success and timely preparation for seizing the next objective. A disadvantage is that the enemy can easily discern the path of advance toward the ultimate objective and therefore prepare and carry out timely counteractions.

Operational Idea

The operational idea (or scheme) is the very core of a design for a major naval operation. In its essence, it is identical to what is in U.S. terms called a "concept of operations" (CONOPS). The operational idea should be developed in a rudimentary form during the naval operational commander's estimate of the situation and then expanded and refined during the planning phase of a major naval operation. Normally, the operational idea for a maritime or land campaign should be developed first, to provide a framework for the operational idea of each subordinate major operation. Otherwise, neither real synergy nor, therefore, the most effective employment of forces is possible. A naval component commander should develop the operational idea pertaining to the employment of naval forces only. If a major naval operation requires substantial participation of air force or land forces in the littorals, the naval operational commander must consider fully the employment of these forces as part of his operational idea.

In general, the operational idea should describe in broad terms what one's naval forces have to do, where, and when. It should express in broad and succinct terms the type of actions by multiservice or multinational forces and the sequence of accomplishing given strategic or operational objectives. The operational idea should be presented graphically on a chart and by operational (vice tactical) symbols.

Operational Sustainment

A major naval operation cannot be successful unless it is logistically supported and sustained in combat in an adequate, reliable, and timely fashion. In general, sustainment is the extension of logistical support from the start of combat actions until the assigned ultimate objective is accomplished. The early identification of critical logistical constraints for the planned major naval operation is indispensable for operational success. Logistically, one's naval maritime force should be as self-contained as possible.

Operational sustainment is required to support one's naval forces throughout all phases of the operation. The primary limiting factors on the operational commander's freedom of action are then the availability of transportation assets and his ability to protect lines of communication from the enemy's attack.

The adequacy of naval operational logistics is measured by its ability to perform its sustainment function and the distribution of supplies.[13] Sustainability is a measure of the capability to maintain logistics support for all users throughout the theater for the duration of the operation.

For example, the U.S. Third Fleet had excellent logistical support and sustainment during the Leyte operation in October–December 1944. Operational sustainment for the Third or Fifth Fleet was the responsibility of the At Sea Logistics Service Group—composed in October 1944 of thirty-four fleet oilers, eleven escort carriers, nineteen destroyers, twenty-six destroyer escorts, and a number of seagoing tugs. This force was divided into ten to twelve task units to fuel the fleet in echelons. A group of nine to twelve tankers was deployed near TF 38, refueling one carrier group after another. About every three or four days, a task unit of three oilers, with escorts, joined this refueling group to relieve those that were almost empty. The latter transferred the remainder of their oil to tankers that were partly full and returned to Ulithi to take on new fuel loads from commercial tankers. Each of these groups was accompanied by an escort carrier carrying replacement aircraft and pilots from Eniwetok, Guam, and Manus in the Admiralties. Combat air patrols and antisubmarine patrols for the oiler groups were conducted by a group of eight escort carriers with their escorts. Rescuing damaged ships was the responsibility of two fleet tugs accompanying each refueling group.[14] TF 38—composed of four carrier groups with nine heavy and eight light carriers, six battleships, three heavy and six light cruisers, three antiaircraft light cruisers, and fifty-eight destroyers—was regularly replenished at sea by TF 30 logistics ships.[15] Fast carrier forces were at sea for thirteen of every sixteen weeks; this was possible only because of excellent logistical organization.[16]

Conclusion

A major naval operation can be successful if the naval operational commanders and their staffs have comprehensive knowledge and understanding of both military and nonmilitary considerations in operational planning. A plan for a major naval operation should be based on a proper determination of the ultimate and intermediate objectives. The operational commander and the planners should not lose sight of the ultimate naval objective. He should avoid selecting multiple intermediate tactical objectives even when forces are superior to the opponent's. Because the time factor cannot be accurately predicted the accomplishment of multiple objectives might require much longer than originally anticipated. This, in turn, can derail the schedule of an entire major naval operation and ultimately slow the operational tempo or even court disaster at the hands of a stronger and more agile opponent.

The ultimate objectives of a major naval operation should be clearly defined and should be achievable with the resources on hand or becoming available. They should not be changed unless the operational situation shifts so drastically that existing objectives should be abandoned or altered. In formulating the ultimate objective, overly broad and ambiguous terms should not be used. In the absence of harmony between one's end and means, ultimate or intermediate objectives should be either changed or scaled down, or additional forces should be assigned.

The design for a major naval operation should focus on the enemy's operational center of gravity. The determination of the wrong center of gravity invariably leads to more losses and the need for additional time to defeat the enemy, and in some cases it can be fatal. Generally, the smaller the margin of superiority, the more important it is to determine properly the enemy's center of gravity and to defeat the enemy quickly, decisively, and with the smallest loss. Otherwise, the pace of the major naval operation will most likely degenerate into attrition warfare. It is also critical to determine the center of gravity of friendly forces, in order to take measures to protect it from enemy action.

The design for a major naval operation is sometimes given short shrift, with almost all the attention given to the design for a campaign. However, a campaign cannot be successful if subordinate major operations are poorly designed, or not designed at all. In addition, without a design for major operations, the tactical employment of forces cannot be effective. The importance of major naval operations is likely to increase in the future, as new technologies will allow smaller forces to accomplish operational and even limited strategic objectives. Many elements in the design for a major naval operation are similar to those in the design for a campaign, with some significant differences due to distinctions in the scale of the ultimate objective.

Warfare, by its very nature, is a series of trade-offs. In each instance, the naval operational commander and staff should properly balance competing demands for scarce resources while still accomplishing assigned operational objectives. Designing a major naval operation is not a simple job amenable to a few hours of discussion. It requires time, imagination, hard work, and above all, sound military thinking and common sense on the part of the naval operational commanders and their staffs. The main purpose of operational design is to make this exhaustive effort a coherent one.

Notes

1. William Lusk Allen, *Anzio: Edge of Disaster* (New York: Elsevier-Dutton, 1978), p. 46.
2. Carlo D'Este, *Fatal Decision: Anzio and the Battle for Rome* (New York: HarperCollins, 1988), p. 96.
3. Fuehrungsakademie der Bundeswehr, *Operative Leitlinie fuer die Fuehrung von Landstreitkraeften* (Hamburg: Fuehrungsakademie der Bundeswehr, August 1992), p. 2.
4. Werner Lange, *Raum, Zeit und Mittel als Faktoren strategischer Zielsetzung. Ueberlegungen zu Napoleons Feldzug in Russland 1812* (Hamburg: Fuehrungsakademie der Bundeswehr, November 1964), p. 1.
5. Ajay Singh, "Time: The New Dimension of Warfare," *Joint Force Quarterly* 4 (Winter 1995–96), p. 56.
6. Max Hastings and Simon Jenkins, *The Battle for the Falklands* (New York: W. W. Norton, 1983), p. 227.
7. Malcolm Llewellyn-Jones, ed., *The Royal Navy and the Mediterranean Convoys: A Naval Staff History* (London/New York: Whitehall History, in association with Routledge, Taylor, and Francis, 2007), p. 131; James J. Sadkovich, *The Italian Navy in World War II* (Westport, Conn.: Greenwood, 1994), p. 289; and J. Caruana, "*Ohio* Must Get Through," *Warship International*, no. 4 (1992), p. 335.
8. Llewellyn-Jones, ed., *Royal Navy and the Mediterranean Convoys*, pp. 81, 129.
9. Ibid., pp. 81, 130.
10. Ufficio Storico Della Marina Militare (USM), *La Marina Italiana Nella Seconda Guerra Mondiale*, vol. 5, *Le Azioni Navali In Mediterraneo Dal 1 Aprile All 8 Settembro 1943* (Rome: Ufficio Storico Della Marina Militare, 1970), p. 355.
11. Ibid., pp. 355–56; Llewellyn-Jones, ed., *Royal Navy and the Mediterranean Convoys*, p. 82.
12. Hermann Franke, ed., *Handbuch der neuzeitlichen Wehrwissenschaften*, vol. 1, *Wehrpolitik und Kriegfuehrung* (Berlin/Leipzig: Walter de Gruyter, 1936), p. 206.
13. Ted O. Kostich, *Operational Sustainment in an Immature Theater: Considerations for Planning and Sustaining a Campaign in a Mid- to High-Intensity Conflict* (Fort Leavenworth, Kans.: School of Advanced Military Studies, U.S. Army Command and General Staff College, 1990), p. 4.
14. Samuel E. Morison, *History of United States Naval Operations in World War II*, vol. 12, *Leyte, June 1944–January 1945* (Boston: Little, Brown, 1958; repr. July 1984), p. 76.
15. Worrall Reed Carter, *Beans, Bullets, and Black Oil* (Washington, D.C.: Department of the Navy, 1953; repr. Newport, R.I.: Naval War College Press, 1998), p. 249.
16. Morison, *Leyte, June 1944–January 1945*, pp. 74–75.

CHAPTER FIVE

The Operational Idea

In general, a sound operational idea for a major naval operation should be simple and flexible. It should be bold, and it should provide for speedy execution. The simpler the operational idea, the better the chances of its successful execution. The operational idea should be sufficiently broad to accommodate changes in the situation in the course of its execution. It should be innovative, avoiding patterns used in the past. It should ensure the decisive employment of forces. It should present the enemy with multidimensional threats that he has little or no chance of countering successfully. It should also surprise and deceive the enemy. The operational idea should be directed at the destruction or neutralization of the enemy's operational center of gravity.[1]

Elements

An operational idea can include a large number of elements common to both a maritime campaign and major naval operation. However, the operational idea for a maritime campaign pertains to the strategic aspects of the situation and that for a major naval operation to the operational situation. Hence, some elements apply only to the design for a major naval operation, not to that for a maritime campaign.

The elements of the operational idea common to a maritime campaign and major naval operation are as follows: selected principles of war, application of sources of power, method for defeating the enemy, the sectors of effort, concentration in the sector of main effort, protection of the friendly center of gravity, anticipation of the point of culmination, deception, sequencing and synchronization, branches, sequels, phasing, pauses, tempo, and reserves. In addition, a major naval operation would also include a point of main attack (defense), operational maneuver and fires, main and supporting forces, tactical maneuver, and regeneration of combat potential. The application of all or even most of these elements would be the antithesis of simplicity and thereby significantly reduce the chances of success in the execution of a major naval operation. Hence, it is incumbent on the part of the naval operational commander and his planners to select only those that will ensure success quickly and with the smallest loss for his forces. In fact, the fewer elements in the operational idea the better. Simplicity of operational ideas is a prerequisite for drafting and executing simple but flexible plans.

Selected Principles of War

An operational idea should provide for the application of the selected principles of war. Not all principles are equally important for success. Hence, the operational commander and the planners should decide which principles of war should be observed and which can be set aside or even violated. Principles of war are rarely clearly stated in an operational idea but are rather implied through selection of the objectives, sectors of effort, deception, and sequencing of actions by one's naval forces.

Some operational ideas have been highly successful although they did not apply some of the most important principles of war. The objective is the most important of all principles of war; its violation would likely cause the failure of the entire operation. For example, the commander in chief of the Combined Fleet, Admiral Isoroku Yamamoto, failed to observe the principle of objective in trying to accomplish several objectives simultaneously in Operation MI (the battle of Midway) in June 1942. He also misapplied the principles of mass and economy of effort. Despite superiority over the much smaller U.S. force, the Japanese suffered a devastating defeat that, in retrospect, signaled the turn of the Pacific War. Likewise, Admiral William F. Halsey violated the principles of objective, economy of effort, and security in his decision to move north in the evening on 24 October 1944 with three fast carrier groups, leaving the San Bernardino Strait unguarded. At the same time, he applied the principle of mass too literally. This was the main reason that Admiral Takeo Kurita, to his great surprise, found that he had free passage to Leyte Gulf. Only Kurita's subsequent decision in the battle off Samar on 25 October to turn away, when on the verge of victory over a much weaker U.S. escort carrier group, saved the U.S. Navy from a humiliating, if largely psychological, defeat.

Sound application of the principles of objective and unity of effort was one of the main reasons for the successful redeployment of two German battle cruisers *(Scharnhorst* and *Gneisenau)* and one heavy cruiser *(Prinz Eugen)* from the naval base at Brest, in Brittany, to Kiel through the English Channel in February 1942 (Operation CERBERUS). The entire operation was brilliantly planned and executed. The Germans achieved complete surprise by timing the passage of their ships though the Strait of Dover during daylight and not during the night, as the British believed they would. The Germans calculated that a transit though the strait during the day would increase their chances of success because the battle cruisers would be better able to defend themselves against the expected massed attacks by the British aircraft and light forces, using shipboard antiair defenses and with fighter cover by Luftwaffe aircraft. The alternative, passing the

Strait of Dover during the night, would require sailing from Brest in the forenoon and steaming though the Channel during broad daylight, thereby giving ample warning to the British. The Germans also maintained an extremely high degree of secrecy in the operation. Only Admiral Otto Ciliax (the commander of the entire force), his chief of staff, and commanding officers of the three heavy ships, and the destroyer force commander knew about the details of the plan for the operation. They also used six code names for the operation to confuse British agents. The Kriegsmarine (German navy) and the Luftwaffe cooperated smoothly, almost flawlessly. In contrast, the British overestimated the flexibility of their airpower while underestimating the high effectiveness of the Luftwaffe. The Royal Navy virtually left the entire operation to the Royal Air Force, and the warships it provided were quite incapable of stopping the German ships. Moreover, cooperation among bomber, fighter, and coastal commands of the RAF was quite poor.[2]

Application of Sources of Power

In designing a major naval operation, the naval operational commander can apply his forces symmetrically or asymmetrically. A *symmetric attack* generally aims to avoid the enemy's strength while attacking his vulnerabilities. It often requires substantial numerical superiority or much more advanced weapons to achieve success and minimize friendly losses or casualties. In contrast, the employment of dissimilar forces can be extremely lethal, especially if the forces attacked are not ready to defend themselves against that threat.

The chances of success are generally greater when acting against the enemy's vulnerabilities than against his strengths. Among the principal advantages of major naval operations in the littorals is the possibility of employing land-based aircraft, and in many cases also ground forces, in an *asymmetric attack*. The employment of land-based air against the enemy fleet or maritime trade is also a form of asymmetric attack. Likewise, the use of ground forces to seize basing areas and ports or of coastal missile/gun batteries against ships are asymmetric methods. In general, major naval operations with participation of other services offer a much wider range of options than those conducted with only naval combat arms. Among other things, multiservice capabilities allow a naval operational commander to employ subordinate forces symmetrically or asymmetrically to generate cumulative effects greater than the sums of their parts.[3] For example, the Japanese, having closely observed and analyzed Allied amphibious landings in the Pacific, changed their method of opposing them from defending beaches to digging in and establishing several defensive lines away from the beaches. In that way, the Japanese countered the superior U.S. firepower and maximized their own advantages. After U.S. troops landed on Okinawa in April 1945 (Operation ICEBERG), the

Japanese offered stubborn resistance in the interior of the island. By the time the last resistance on the island ended in late June 1945, the Japanese had lost 110,000 men in combat, but they had inflicted very heavy losses on the attacker: American battle casualties numbered about forty-nine thousand, including some 12,500 killed or missing.[4]

Sequential versus Simultaneous Attacks

The operational idea should clearly describe how available combat potential is applied in terms of time—sequentially, simultaneously, or combined. Which method will be employed depends on the content of the ultimate operational objective and political, legal, and other nonmilitary limitations imposed on the use of military forces. The most decisive and usually quickest results are achieved by simultaneous application of overwhelming force. The incremental use of power has invariably proved less effective or even highly detrimental, whether it is used against a strong enemy or an inferior one. Such an approach gives a resourceful enemy time to recover from blows. It usually prolongs the duration of a major naval operation and therefore results in unnecessary losses for friendly forces.

Method for Defeating the Enemy

A sound operational idea should in broad terms describe methods for defeating the enemy. In general, while the majority of effort should be aimed at destroying or neutralizing the enemy's center of gravity, other elements of enemy power must be attacked as well. Obviously, methods selected for destroying the enemy's center of gravity should be consonant with the operational objective to be accomplished.

The operational idea should contemplate direct or indirect action to destroy or neutralize the enemy's operational center of gravity. Normally, if the naval operational commander has overwhelming forces available, the quickest and most effective approach is to attack a major part of the enemy's forces directly. "Direct attack" means attack by one's strengths against the enemy's strengths.

In a major amphibious operation, all efforts should normally be directed to defeat or neutralize quickly the enemy's operational center of gravity. However, if the attacker is inferior in some critical aspect of combat power or is unable to ensure logistical sustainment in the early phase of the operation, he will have initially to seize certain physical objectives, such as key ports/airfields, before he can attack the enemy's operational center of gravity, directly or indirectly. For example, in planning their invasion of Norway (WESERUEBUNG NORD) the Germans focused on seizing selected ports and airfields before attacking the enemy's center of gravity—the Norwegian troops defending the capital of Oslo and the forces defending the central and northern parts of Norway.

In a major naval operation, the use of feints or deception to influence the enemy's naval operational commander to move his forces and operational center of gravity from a certain sea or land area and thereby uncover other critical strengths can represent an indirect attack upon them.

Decisive Points and Center of Gravity

The enemy's operational center of gravity can be defeated or neutralized by using physical, cybernetic, or informational sources of power to attack a series of the enemy's decisive points. A decisive point may or may not have readily identifiable physical characteristics. Determining and attacking an incorrect decisive point may prove fatal to the design and execution of the basic plan, even if the correct center of gravity has been identified.[5]

Decisive points can be used as keys to the enemy's center of gravity. The destruction, annihilation, neutralization, or serious degradation of selected decisive points may generate a "cascading" effect in other parts of the outer core of the center of gravity. In some cases, by attacking decisive points one can defeat the center of gravity without having to attack it directly.[6]

If the enemy's center of gravity is difficult to attack directly, because it is physically strong or its essential elements are well protected or concealed, the indirect method of attack should be used. This method is also often suitable for degrading or neutralizing intangible elements of the enemy's center of gravity. Indirect methods of destroying or neutralizing the enemy's center of gravity are aimed at identifying and then exploiting critical vulnerabilities, sometimes even critical strengths. An enemy's vulnerability is most valuable as a target when its destruction contributes to the elimination or serious degradation of the enemy's center of gravity. Such vulnerabilities are found in the enemy's physical, moral, and cybernetic domains. In an indirect approach against the enemy's center of gravity, the naval operational commander should seek to attack the enemy force from an unexpected direction or at an unexpected time, thereby avoiding the enemy's strengths by exploiting his vulnerabilities.

The naval operational commander and planners should identify all potential decisive points, to determine which of them offer the best options for eventual attack on the enemy's center of gravity. In general, a decisive point can be located in the physical proximity of the center of gravity or some distance away. For instance, a decisive point in the enemy's computer network can be located on land or in space several hundred or thousands of miles from the enemy's center of gravity.

A decisive point is usually an enemy's critical vulnerability, but it can also be one of the enemy's sources of strength, even a critical strength. If such a decisive point is exclusively or predominantly physical in nature (a commanding height, railroad/road junction, port/naval base, airport/airfield, C4 node, supply depot, computer "server farm," etc.), it can be subjected to direct physical attack. For example, in the initial phase of the German invasion of Denmark in April 1940, six ports intended for debarking troops (Copenhagen, Korsor, Nyborg, Middelfart, Esbjerg, and Thyboron) were decisive points for the invasion planners, as were seven Norwegian ports (Oslo, Kristiansand, Arendal, Stavanger, Bergen, Trondheim, and Narvik). In the Allied invasion of Guadalcanal in August 1942, islands near Guadalcanal (Tulagi, Florida, Russells, Malaita, San Cristobal, etc.) and the sea passages between could be considered geographically oriented decisive points; the Japanese airfield on Guadalcanal, the seaplane base at Tulagi, and their air defenses could be considered force-oriented decisive points. In the 1950 UN landing at Inchon, the oyster-shaped island of Wolmi-do was in operational terms a decisive point, because it controlled the approaches to the port of Inchon, only some eight hundred yards away. The island was connected to the port by a narrow causeway.[7]

For a naval force at sea, a decisive point can be a single ship or a group of underway-replenishment ships. For example in the battle of the Coral Sea in May 1942, the U.S. fueling task group (TG 17.2), composed of two tankers *(Neosho, Tippecanoe)* and two destroyers *(Sims, Worden)*, should have been a decisive point for the Japanese planners.[8] The destruction of one or both tankers would have had a major impact on the ability of the U.S. fast carrier group *(Yorktown, Lexington)* and accompanying cruisers and destroyers of Task Force 17 (TF 17) to sustain their operations against the Japanese carrier force. For a carrier strike group or expeditionary strike group today, the decisive point can be a nuclear submarine or large surface combatant upon which defense against an enemy submarine or early warning and defense against enemy aircraft would be based.

If the enemy relies extensively on computerized systems for command and control, or intelligence, it is possible to attack the enemy's center of gravity indirectly through an attack against cybernetic-oriented decisive points. Computer networks are in many ways similar to transportation networks. The planners might select a method of attack that focuses on critical or key nodes whose destruction or annihilation would have a ripple effect on the functioning of the network as a whole.

Protection of the Friendly Operational Center of Gravity

A sound operational idea for a major naval operation includes details on forces and the method of protecting a friendly operational center of gravity. The key here is to know and fully understand the enemy's perspective on what the friendly center of gravity is, because the enemy might use a different approach or method to determine it.[9]

Optimally, the naval operational commander should have sufficient forces to ensure full protection of all the forces deployed in a maritime theater, including coastal installations/facilities. However, this is rarely the case, because resources are not unlimited. Therefore, the naval operational commander should balance competing demands for resources so as to accomplish the principal objectives while also having sufficient forces to conduct operational fires and provide operational protection.

The naval operational commander should do everything possible to provide sufficient protection for the friendly center of gravity while avoiding the danger of overprotecting the wrong center of gravity.[10] In general, a highly capable but not excessively strong force should be assigned to protecting the friendly center of gravity; otherwise, one could be exposed to a devastating attack. In historical cases when operational commanders and planners have included protection of their centers of gravity as part of the operational idea, major naval operations have usually been successful. In contrast, many major operations have failed or only partially succeeded because the commander paid inadequate attention to protecting his own center of gravity in the course of pursuing a beaten enemy force.

For example, during the battle for Leyte in October 1944, the Japanese failed to protect adequately the operational center of gravity of their surface forces—the 1st Diversionary Attack Force, under Vice Admiral Takeo Kurita—during its movement from Borneo via the Sibuyan Sea and then through the San Bernardino Strait southward to the Tacloban anchorage. Kurita sent several urgent requests for support to Vice Admiral Shigeru Fukudome, commander of the Japanese land-based aircraft—the Fifth and Sixth base air forces—on Luzon. Fukudome repeatedly declined these calls for help. He later justified his decision on the basis of the inherent difficulties for land-based aircraft in providing air cover to a constantly moving surface force. In his view, sending fighter aircraft and scouting planes to search for enemy submarines would seriously weaken the offensive power of his forces. He contended that the best protection he could provide to Kurita's force was to concentrate the entire strength of the base air forces on attacking enemy carrier forces then deployed east off Luzon. However, owing partly to bad weather, these attacks were not successful; base air force aircraft attacked two carrier groups, hitting and heavily damaging the light carrier *Princeton*.[11] After the war, Admiral Soemu Toyoda, then commander in chief of the Combined Fleet, explained that one reason the base air forces failed to provide adequate protection to Kurita's force was that their strength had been sorely depleted after the air battles off Formosa and for the Philippines. However, in his view, the main reason was the "lack of unity between tactical and operational thinking." In addition, the base air forces and surface forces had not been well enough trained in cooperation.[12] However, Toyoda, as the fleet commander, had not only the authority but also the responsibility to order Fukudome to cooperate.

Sectors of Effort

In designing a major naval operation, the operational commander and his staff have to decide early on which part of the theater or area of operations will be the *sector of main effort (thrust)* and which will be the sectors of secondary effort.[13] A sector of main effort is usually selected in a sea/ocean area where the principal objective must be accomplished. Often, the enemy will deploy or concentrate his main forces in that sector.

Designation of a sector of main effort is critical in focusing efforts to destroy or neutralize the enemy's operational center of gravity. By designating the proper sector of main effort, the operational commander tells his planners which friendly forces should receive the greatest attention in planning fires and logistical sustainment. Prior to the beginning of a major naval operation, the sector of main effort should be kept secret from the enemy.

In a major naval operation, any serious error in determining the enemy's proper center of gravity will result in a wrongly selected sector of main effort. If the error is corrected in time, the operational commander should *shift the sector of main effort* to compensate for the error. This shift can also be caused by a change of plans due to some unexpected opportunity to inflict higher losses on the enemy. A shift in the sector of main effort would normally require a change in support priorities among maritime forces. Thus, flexibility in shifting the main effort should be an integral part of the operational design.[14] For example, on 9 December 1941 Admiral Karl Doenitz requested permission from the German naval high command to deploy twelve 1,100-ton Type IXC U-boats off the American east coast shortly after the United States entered the war. This request was reduced to only six U-boats, and of these, three large U-boats were directed to the Mediterranean and the others to patrol the area between Gibraltar and the Azores. Nevertheless, by mid-January 1942 Doenitz was able to concentrate seven smaller Type VII (750 tons) off Newfoundland.[15] These U-boats mounted an offensive against Allied shipping off the East Coast in February 1942.

The sector of main effort can also be shifted because of unexpectedly strong resistance. For instance, the Allied threat from the air forced the Germans to shift the sector of main effort of U-boats operating in the northern Atlantic. In the second phase (July 1940–March 1941) of the Battle of the Atlantic, the sector of main effort of the U-boats shifted to the Western Approaches. There was no discernible sector of main effort in the third phase (April–December 1941); U-boat sinkings of Allied merchant ships were spread over most of the northern Atlantic and the eastern part of the central Atlantic. After August 1942 and until May 1943 (Phase V), the great majority of the Allied ships were sunk in the central part of the northern Atlantic. Afterward and until the end of war, the U-boats were unable to mass their attacks on the Allied shipping even in a relatively smaller part of the northern Atlantic, because they were essentially defeated by the Allied antisubmarine (ASW) defenses.

The *sector of secondary effort* encompasses the area in which the enemy's supporting forces are expected to operate. It is characterized usually (but not always) by smaller depth and width than the sector of main effort. Actions in the sector of main effort can hardly be successful unless supported by other friendly forces in a sector of secondary effort. For example, in a major conflict in an enclosed-sea theater like the Persian (Arabian) Gulf, the Baltic, or the Black Sea, the sea's only exit and its approaches would most likely be initially the sector of main effort for both the attacker and the defender, while the remaining part of the sea would be the sector of secondary effort.

Point of Main Attack (Defense)

An important element of the operational idea is selection of the *point of main attack* (or *point of main defense*)—a part of the sector of main effort where most of the important actions of one's main forces will take place. By properly determining the point of main attack, the attacker (or defender) can obtain relative superiority over the enemy even if the overall ratio of forces is unfavorable. Good tactical and operational intelligence is critical in the selection of the point of main attack. Effective employment at the point of main attack also requires a deception plan.

Points of main attack (defense) for a major naval operation are often difficult to determine, because of the inherent fluidity and rapid changeableness in a situation at sea. The exception is an amphibious or anti-amphibious landing operation. In that case the point of main attack (or defense) is usually selected among decisive points offering an indirect approach for attacking the enemy's center of gravity. For an amphibious landing, the point of main attack is usually a part of a coast or a beach that is the least defended and at the same time offers the best opportunities to advance inland or that offers the most favorable conditions for attacking the enemy's center of gravity indirectly, or a port of a gulf/bay whose capture is critical for logistical sustainment.

In a major naval operation aimed to interrupt the enemy's maritime commerce, the selection of the point of main attack is usually influenced exclusively by shipping density in a certain sea or ocean area, such as the focal point of shipping or choke point—

specifically, approaches to larger ports, channels, narrows, and international straits or canals. For example, in the German campaign to cut off Allied trade in the Atlantic in 1939–45, points of main attack for U-boats included the English Channel and its approaches, the Bristol Channel, and the Strait of Gibraltar. In other cases, large ports or shipyards and ship-repair facilities whose destruction would cripple maritime trade in a certain area could be selected as points of main attack. For the defender, such points should obviously be considered main points of defense.

Main versus Supporting Forces

The sectors of effort also dictate where the main or principal forces and supporting forces should be concentrated or employed in a major naval operation. Normally the actions of naval forces will be focused on the enemy's main forces, in the sector of main effort, whose destruction or neutralization would lead to accomplishment of the main objective of the entire operation. To achieve success in the main sector, supporting forces should be capable of destroying or neutralizing the enemy's covering or supporting forces, supporting groups, or patrol forces to prevent them from interfering with friendly main forces in the main sector of effort.[16] Another purpose of supporting actions is to neutralize enemy ASW in the theater, weaken the enemy's theaterwide air defenses, and destroy or neutralize his C4 nodes.[17] These actions would be usually conducted simultaneously with the actions of main forces in the sector of main effort.

The greatest problem is in balancing the size/mix of forces in the sector of main effort and of those in the sector (or sectors) of secondary effort. In general, a more mobile and lethal force allows more flexibility in shifting the sector of main effort. Allocation of combat and combat-support forces among the sectors of main and secondary effort is an integral part of the commander's estimate process.[18] This is more an art than a science. It calls for good judgment and experience on the part of the operational commander. Sometimes the sectors of effort are properly selected but the distribution of forces among them is not. For instance, in the Japanese major naval operation (MI) that led to the battle of Midway on 4–6 June 1942, the sector of main effort was a rather large part of the surface, subsurface, and airspace around the island of Midway, while the Aleutians (Operation AL) represented the sector of secondary effort. However, the Japanese assigned too large a force for a diversionary effort in the north instead of strengthening their forces in the sector of main effort at Midway.

Operational Maneuver and Fires

The operational idea for a major offensive naval operation will usually include operational maneuver. An *operational maneuver* is conducted within the boundaries of a given maritime theater of operations and is aimed at exerting decisive impact on the outcome of the individual major naval operations in a maritime campaign, either by securing the advantage offered by some particular position before attacking the enemy's operational center of gravity or by striking at some critical vulnerability. In either case, the intent is to force the enemy to react operationally or even strategically. The outcome of a maritime campaign might depend on the success or failure of the operational maneuver for the initial major naval operations.

Operational maneuver is planned in connection with major naval operations aimed at destroying the enemy fleet at sea or its bases or at attacking coastal installations/facilities, and for major amphibious landing operations. An example of operational maneuver at sea is the movement, in an amphibious landing operation, of amphibious forces and of forces of close and distant (operational) cover from the initial base of operations to the line or area from which landing craft or helicopters would move inland. For example, forces that took part in the Allied amphibious landing on Leyte in October 1944 (Operation KING II) conducted an operational maneuver by eight separate elements from staging areas at Hollandia, New Guinea, and Manus, in the Admiralties, to positions off Suluan, Homonhon, and Dinagat islands fronting the landing beaches.

Concentration in the Sector of Main Effort

The operational idea should provide for quick and, if possible, covert concentration of one's naval forces in a sector of main effort and at the point of main attack (defense). In general, concealed and rapid concentration of naval forces is critical for the successful outcome of the initial tactical actions. It is especially critical for the success of numerically inferior forces. The larger a naval force, the more critical is timely concentration. Concentration of naval forces is usually conducted in a sea/ocean area open to enemy attack or in areas where control is in dispute.[19]

In the past, a naval force was considered concentrated when all its elements were within supporting distance of each other—in other words, when adjacent elements of the force could apply their full strength in coordination against the parts of the enemy force they respectively opposed.[20] However, the concept of concentration should not be interpreted literally. Today, forces, especially if they are netted, do not necessarily need to be massed physically in a certain area to be "concentrated." The high speed, precision, and ever-longer range of weapons and surveillance sensors now allow a force to be "massed" even when its various parts are dispersed over a large part of the theater. What counts most is "massed effect." For example, on the open ocean or in the littoral waters fronting the open ocean, netted forces can be geographically widely separated but still capable of generating massed effect by virtue of the long ranges of their weapons and sensors.

Depending on its purpose and scope, a concentration can be tactical, operational, or strategic. *Tactical concentration* is aimed at massing naval forces prior to a naval tactical action, and it takes place in a specified naval combat zone or sector. *Operational concentration* is planned and conducted as a part of a major naval/joint operation or maritime campaign. Normally, it takes place in a given maritime area of operations. *Strategic concentration* is aimed at creating superior strength in a particular maritime theater of operations or theater of war. It pertains to concentration of the entire fleet, or of its major part.

Operational concentration of naval forces can take place off or on the battlefield. When force elements are brought within supporting distance *before* arriving in the area of operations, they are said to have been *concentrated off the battlefield.* If they are grouped within a supporting distance as the combat action is about to start or shortly after it has begun, they have been *concentrated on the battlefield.* In a war at sea, concentration off the battlefield is most often used in amphibious landing operations. The amphibious task force, forces providing direct support, and those in distant cover and support sail from their staging areas, within mutually supporting distance, to the assigned debarkation areas in the amphibious objective area (AOA).

If various elements of one's naval forces would not arrive within supporting distance until they actually become engaged with the enemy, their concentration is carried out in the form of a maneuver. Concentration on the battlefield generally enhances the factor of surprise. It also makes it difficult for the enemy to deduce where the next blow will come from until it is too late to react. However, such a concentration requires meticulous planning and coordination of the movement of naval forces and aircraft. Also, the opponent operating from a central position has the opportunity to defeat converging forces in detail, especially if the advance of elements of the converging forces is delayed by faulty timing, bad weather, or some other reason. Concentration on the battlefield is often used to defeat a large part of the opposing fleet. This was the case in the Japanese invasion of the Netherlands East Indies in 1942. The Japanese Combined Fleet also attempted, but failed, to carry out concentration on the battlefield in its ill-fated operations in the defense of the Philippines in October 1944.

Tactical Maneuver

Closely related to the operational maneuver and concentration at the point of main attack is tactical maneuver—a movement of individual force elements toward assigned attacking positions within the effective ranges of their weapons. This type of maneuver is an element of the operational idea for a major naval operation aimed at destroying the enemy fleet at sea or its bases and for one aimed to destroy merchant shipping at sea. In a major amphibious landing operation, tactical maneuver encompasses the movement of amphibious forces from their line of departure to the landing beaches. Their maneuver is usually supported by tactical fires of surface ships and aircraft.

The key considerations for planning a tactical maneuver include the number of force elements, selection of the basic and alternate routes, cruising formation, method and the force required for conducting maneuver, the sequence of actions of the individual force elements, measures of combat support, and measures for cover and concealment. The naval operational commander and planners should select a cruising formation most favorable for the use of weapons, force protection, and mutual support. This is especially critical for the success of forces assigned to the sector of main effort.[21] In littoral waters, one of the preconditions for the successful execution of tactical maneuver is obtaining and maintaining basing/deployment control. Tactical maneuver should be conducted, if possible, during the night or in bad weather. Coastal defenses should be alerted to provide fire support to friendly forces during the execution of tactical maneuver in littoral waters.

Operational Deception

A sound design for a major naval operation should include a plausible plan for operational deception. In general, deception is aimed at misleading the enemy about one's real intentions, future decisions, and friendly courses of action. It can be designed to confuse and disorient the enemy about the time and place of attack, thereby achieving surprise. Deception can also be designed to create an illusion of strength where weakness exists or to paint a picture of weakness where strength exists. It can be intended to induce the enemy to direct attention and concentrate forces in the wrong place and thereby violate the principle of concentration of forces in terms of the factor of space.[22] Another purpose of deception can be to cause the opponent to waste time by concentrating his forces at the wrong time and, preferably, on nonexistent objectives.[23]

A friendly center of gravity can be protected much more effectively if deception is used to conceal, or mislead the enemy about, one's true capabilities or types of forces or the location of the center of gravity. Deception can be used to mislead an opponent about the deceiver's capabilities, either by creating an exaggerated perception of the numerical strength and quality of forces or by concealing the locations of forces. It is often intended to surprise the enemy and thereby considerably enhance the success of forces in the initial phase of a major naval/joint operation.

Deception can be intended to overload the enemy's intelligence collection and analytical capabilities or to deny him vital information. This, in turn, can prevent the enemy from forming an accurate and timely picture of an operational situation. Successful deception invariably introduces "noise" into enemy intelligence collection and analytical work and weakens the clarity of the signals received.

In the strict definition of the term, *operational deception* pertains to actions and measures to deceive the enemy as to the time, place, and details of a planned major operation conducted as part of a campaign or a major joint/combined operation with a strategic objective. Successful operational deception protects the operational commander's intent from the enemy's intelligence-gathering. It can reinforce enemy expectations and preconceptions about friendly force.

Actions of one's naval forces might surprise the enemy if the deceiver creates an impression of routine activities by gradually conditioning the enemy to a certain repetitive pattern of behavior. This method was used by the Germans in preparing the operational redeployment of two battle cruisers *(Scharnhorst, Gneisenau)* and one heavy cruiser *(Prinz Eugen)* from Brest through the English Channel in February 1942. The Germans increased the intensity of their radar jamming gradually over time so that the British became acclimated to it and did not realize that it eventually became so intense that British radar was almost useless.[24]

Operational deception should be coordinated with strategic deception. Any disconnect between these two invariably creates difficulties in their executions and could even endanger the success of the entire operation. For example, the deception plan (code-named BARCLAY) in support of the Allied landing on Sicily (Operation HUSKY) depended on the successful execution of a deception plan known as MINCEMEAT. The actual organization of MINCEMEAT was vested in a British intelligence group known as "Force A" in Cairo, which sent a team to Eisenhower's headquarters in Algiers as a deception planning staff.[25] The general objective of BARCLAY was to induce the Germans and Italians to maintain and reinforce sizable forces in the Balkans and in southern France, thereby limiting their ability to reinforce Sicily.[26] The original broad outline of that plan envisaged "disclosing" to the enemy Allied intentions for (fictional) attacks by the notional British Twelfth Army between 26 May and 6 June on western Crete, the Peloponnesus, and the islands of Pantelleria and Lampedusa, and by the U.S. Seventh Army on Sardinia and Corsica and in southern France. However, on 15 May, Force A decided to postpone all these fictional assaults, leak that news to the Germans and Italians in late May, and revive the same threats two months later. The revised BARCLAY involved the same sequence of events, but these notional attacks would now peak about twelve days after the actual landing on Sicily (10 July); eleven days later, the fictional assault on southern France would take place.[27]

The Allies also planned a naval feint, code-named Operation FRACTURE, as part of the overall deception effort for the landing on Sicily. This action was planned for the early morning of D-2 (i.e., two days before the actual landing) and for the night between D-3 and D-2 in the area of Marsala–Mazzara. The Allied forces assigned to FRACTURE consisted of one battleship and cruiser division. In addition, a force of coastal craft would present the enemy defenders with the threat of landing by bombarding the town of Marsala and the island of Marettimo and conducting a demonstration off Marsala.[28]

The Allied planners tried to convince Hitler and OKW (the high command of the German armed forces) that a landing on Sicily was merely a preliminary to, and subordinate part of, a major Allied effort that would include the capture of Sardinia and Corsica and the invasion of the southern Balkans.[29] As it turned out, the Germans sent substantial forces to the Balkans. They also divided their forces among Sardinia, Sicily, Corsica, and the mainland of Italy. Yet not all Axis commanders were deceived; sufficient signs—such as a concentration of Allied landing craft and of ground troops in North African ports, as well as increased air activity—unmistakably pointed to Sicily as the next Allied objective.[30] Italian intelligence was accurate in its assessment on 24 June that the Allied exercise near Oran was a rehearsal for a paratroop drop on Sicily. A week later another Italian intelligence report correctly predicted that the next Allied objective would be Sardinia or Sicily, most likely Sicily. However, the authorities in Rome and Berlin did not believe these reports, and so the Allied landing on Sicily achieved complete operational surprise. In addition, naval tactical feints kept some Axis forces away from the landing beaches on Sicily. The Allies were also successful in making the defenders believe that their amphibious convoys were moving toward Greece, because in fact the convoys' routes converged in the general area of Malta; then, however, they turned and approached Sicily under the cover of darkness.[31]

Sometimes operational deception has to be maintained after the initial major operation in a new campaign has been executed but while forces are still threatened by enemy forces within striking distance. In that case, the initial deception plan must be modified or completely rewritten. The Allies were faced with such a situation after their landing in Normandy on 6 June 1944. Because of stiff German resistance, the lodgment on the continent was tenuous for several weeks after the landing. The Allies activated deception plan FORTITUDE SOUTH II, aimed to persuade Hitler and General Gerd von Rundstedt, CINC of the West, that there was a continuing Allied threat to the Pas de Calais area. This threat was maintained by a large notional force, composed of the 1st U.S. Army Group and U.S. XIX Tactical Air Command, on the eastern and southeastern coasts of England. The threat was substantiated by considerable radio activity and about 250 dummy LCTs (landing craft, tank) deployed in the harbors and estuaries of eastern and southeastern England.[32] FORTITUDE SOUTH II used forces similar in size to those used in FORTITUDE SOUTH itself. The big difference was that notional forces were used instead of real ones. The fresh U.S. troops had to be sent as reinforcements to Normandy. Notional forces were composed of two divisions from the U.S. Fourteenth

Army and one division from the British Fourth Army. These notional assault formations also "carried out" intensive amphibious exercises.[33]

Operational Sequencing

One of the main preconditions for the successful execution of a major naval operation is sound sequencing of intermediate objectives. The main purpose then is to arrange events in such a progression that collectively they lead to the accomplishment of the assigned ultimate objective in the shortest time possible and with the least loss for friendly forces. In the naval context, *tactical sequencing* is conducted for naval battles, engagements, strikes, and certain other naval tactical actions. It determines the objectives and tasks to be accomplished by one or several combat arms/branches of a single service, or two or more services. *Strategic sequencing* deals with sequencing campaigns or a series of major operations as part of a campaign in progress in two or more theaters of operations, within a single theater of war, or in adjacent or widely separated theaters of war.

Operational sequencing is aimed at arranging the sequence of major tactical or operational objectives/tasks in the course of a major naval operation. It can be planned for each major phase—from mobilization, to strategic or operational deployment, to combat, posthostilities, redeployment, and demobilization. However, the focus is normally on the sequencing of events during combat-force employment and the employment of all or selected operational functions in support of a major operation or campaign. Inadequacy in any element of operational functions, if left unresolved, can cause premature culmination, endangering the success of the entire maritime campaign or major naval operation.

Operational sequencing provides the path for synchronization of combat forces in accomplishing each of these intermediate objectives. Another purpose of sequencing is to develop a force list for the forthcoming operation by calculating how many and what types of forces are required to accomplish each intermediate objective or main/partial (component) task and then identifying the forces' shortfalls. Task organization should be based on the force list derived from combat-force sequencing.

In general, operational sequencing can be arranged in terms of objectives, tasks, time, or a combination of these. *Objective-oriented sequencing* is designed to concentrate friendly forces selectively in time and space, to create conditions that will translate operational or tactical success into operational or strategic advantage. The planning should start with determining the ultimate objective of a major naval/joint operation and then identify the required series of intermediate objectives.[34] In war at sea, intermediate objectives are often the enemy force or force-related elements. The linkage between an objective and the options for accomplishing it provides a framework by which planners can sequence a series of events.[35] However, there is an inherent tension in trying to sequence events based on a complex interaction among operational factors (time-space-force) and the assigned objectives.

Task-oriented operational sequencing refers to the order in which the main tasks are accomplished. Care in sequencing offers flexibility during execution, because main tasks can be modified or canceled if unanticipated events occur, but the ultimate objective would remain unchanged until it is accomplished. A drawback of this method is its complexity, in that for an operational objective a number of main tasks (each of which consists of a number of partial or component tasks) must be carried out. Too many tasks will invariably lower the operational tempo and possibly fragment the effort. On the other hand, too few tasks would not accomplish the ultimate objective of the operation.

Generally, operational sequencing should not encompass tactical objectives/tasks, either during the planning process or, even worse, in the course of execution of the operation. Not only would their inclusion unnecessarily complicate the entire process, but they might in fact become more important than the operational objectives and even derail the entire sequencing timetable. Among other things, this was a major reason for the Japanese failure in Operation MO in May 1942. The Japanese striking force (fast carriers) was tasked to ferry eighteen Zero fighter aircraft from Truk to Rabaul. However, unforeseen events caused this simple additional task to disrupt the timetable for the entire operation.[36] Everything went according to plan until 2 May, when the striking force reached a position about 240 miles northeast of Rabaul. That day and the next, the Japanese commander tried to launch nine Zeros to Rabaul; both attempts failed because of bad weather, which also prevented the Japanese carriers from refueling at sea. Thus, precious time was lost and the entire sequencing schedule was affected; moreover, the Japanese carrier force could not get within range to protect Tulagi until 5 May, too late to have any real impact.[37]

Means should always match the ends; otherwise, no objective or task can be accomplished easily, and some cannot be done at all. By understanding relationships between ends and means, planners can decide whether events should occur sequentially or simultaneously.[38] Also, the scale of the planned major naval operation should be in harmony with the means available. The basic sequencing plan should be adhered to unless some extraordinary event requires a change in the sequencing schedule. A change in the original timetable might be warranted if the enemy abandons his positions or areas, if a new intelligence estimate shows a drastic change in the operational situation, if the enemy's resistance is unexpectedly weak or surprisingly strong, if attrition of friendly forces is lower or higher than expected, or for some other reason.

Operational Synchronization

Synchronization is another critical element in designing a major naval operation. Its success is contingent on the sound sequencing of operational or major tactical objectives. Improper sequencing of objectives/tasks usually dooms synchronization. In addition, various elements of combat support should be synchronized with the actions of combat forces; otherwise, a major naval operation might fail.

Tactical, strategic, and operational synchronization are differentiated by scale. *Tactical synchronization* is applied in planning and executing naval tactical actions. Sound and effective synchronization of forces at the tactical level is one of the prerequisites for the success of operational synchronization. *Strategic synchronization* deals with the employment of military and nonmilitary sources of power at a theater-strategic or national-strategic level. It is applied in the conduct of two or more campaigns in a given theater of war or two adjacent ones.

Operational synchronization pertains to arranging actions in time, space, and purpose for the planning and execution of a major naval operation. It is much more complex than tactical synchronization, because of the much larger factors of space, time, and force. The naval operational commander needs to synchronize combat support during the major phases of combat-force employment. Synchronization involves several simultaneous or sequential strikes or attacks against targets in the sector of main effort; the aim is to neutralize in a timely way the enemy's ability to interfere with actions of friendly forces in the main sector. These actions may include sweeping certain sea or ocean areas for mines, deploying submarines, or laying mines to restrict the maneuver of the enemy's forces. Precise synchronization in terms of space and time is needed between strikes by the main forces and supporting actions aimed to create favorable conditions for the main task.[39]

In their invasion of Norway in April 1940, the Germans developed a plan for synchronized employment of all three services of the Wehrmacht. The plan envisaged a quick occupation of the Jutland Peninsula and the island of Fuenen by surprise, followed by the occupation of Zealand. In Norway the first task was to seize important points on the coast by surprise landings from the sea and air. The plan was daring and original. The troops were to be transported to several points by destroyers; artillery, equipment, and supplies were to be carried in by transports disguised as cargo ships. Airborne troops were to be used in Oslo and Stavanger. Almost the entire German surface fleet, divided into seven groups, was required for the operation. The naval part of the initial operation envisaged that synchronized and covert movement of seven "export echelons"—warships and merchant ships with troops and artillery—would seize control of seven ports in Norway as follows:

- One group of ten destroyers carrying two thousand troops of the 3rd Mountain Division to Narvik would be accompanied by two battle cruisers *(Gneisenau* and *Scharnhorst),* which would break off at an agreed point near their destination and create a diversion in the Arctic.
- A group carrying 1,700 troops and escorted by one heavy cruiser and four destroyers would reach Trondheim.
- Some 1,900 troops embarked on two light cruisers, the auxiliaries, two torpedo craft, and five fast patrol craft would also head for Trondheim.
- A group consisting of 1,100 troops carried by one light cruiser and auxiliary ship, three torpedo craft, and seven fast patrol craft would move to Kristiansand and Arendal.
- About two thousand troops escorted by one pocket battleship, a heavy cruiser and a light cruiser, three torpedo craft, two armed whalers, and eight minesweepers would disembark at Oslo.
- Some 150 troops on board four minesweepers would head for Egersund.

A "tanker echelon" of eight ships carrying fuel for Narvik, Trondheim, and Stavanger would sail ahead of the export echelons. The arrival of all groups was so timed that they would reach their assigned positions off Norway's ports at "W[eser] hour." Another group of fifteen transports of the 1st Sea Transport Echelon would arrive at Oslo, Kristiansand, Bergen, and Stavanger later in the day (9 April). Subsequent transport echelons would be directed only to Oslo.[40]

In another example, the Japanese synchronization plan for the SHO-1 plan that led to the battle of Leyte was deeply flawed. The planners envisaged the 1st Diversionary Attack Force sailing out from Lingga Archipelago (southeast of Singapore), advancing via the San Bernardino Strait toward Leyte Gulf while destroying any enemy surface forces opposing it, and then attacking the Allied transports in the gulf. In coordination with this attack, the Mobile Force was to lure the enemy north and attack elements of his forces at the most favorable opportunity. The Second Diversionary Attack Force would come under the command of Southwest Area Fleet (with headquarters in Manila) and would cooperate with the army in conducting a counterlanding on the western coast of Leyte. The Base Air Force would concentrate in the Philippines for an all-out attack on the U.S. fast carrier groups. The Advance Force (submarines) would attack damaged enemy vessels and amphibious convoys with all the forces under its command. The First Diversionary Attack Force would execute its attack against the enemy invasion forces at the landing point on "X-day." The Mobile Force would advance to the area east of Luzon on X-1 or X-2.[41]

The main problem with the Japanese synchronization plan was that its success depended on precise coordination, which is difficult to achieve even under the best of circumstances. For example, it was essential that Admiral Soemu Toyoda, the commander in chief of the Combined Fleet have information on the movements of the Allied forces five to six days in advance in order to have sufficient time to coordinate the movements of the Mobile Force and First Diversionary Attack Force. However, even this advance warning would not have been enough, because by then the Allied transports would have been unloaded in the Tacloban anchorage. Surprisingly, the Japanese did not station their submarines off the Allied ships' most likely departure points from Manus and Hollandia, so they missed the opportunity to obtain additional warning time; these bases were also inadequately reconnoitered by Japanese land-based aircraft.[42] In addition, the space factor adversely affected the synchronization plan. The distance between the Bungo Suido (one of the two southern exits of the Inland Sea) and Singapore is about 2,500 miles, and the Philippines lie approximately halfway between these two points. Moreover, the Japanese should have known that their perennially poor communications would not allow synchronization of the movements of their far-flung forces. Clearly, the synchronization schedule was too rigid, because it made no provision for unforeseen events, such as U.S. submarine or carrier aircraft attacks on the Japanese surface forces in Philippine waters. Kurita's First Diversionary Attack Force lost about four and a half hours avoiding U.S. carrier aircraft attacks in the Sibuyan Sea; this delay by itself derailed the initial synchronization schedule.

Branches

The operational commander should not be bound to a single course of action in planning for a major naval/joint operation. Sufficient flexibility should be built into the basic plan to preserve freedom of action. The naval operational commander accomplishes this by anticipating and planning *branches*—options built into the basic operation plan for a major naval/joint operation. They are contingencies within a given phase of the operation.[43] Branches allow the naval operational commander to anticipate future enemy actions that might lead to the need to modify his basic plan drastically. Such a situation could arise from unexpected action by the enemy, nonavailability of forces, or even a change in the weather within the operational area. As conditions change, the naval operational commander should be prepared to maintain the continuity of the operation by redeploying other friendly forces in the theater.

Sequels

In designing a major naval operation, planners need to envisage one or more successive major or minor tactical actions—known as *sequels*—after an intermediate objective is accomplished. Sequels and branches are related but not identical in their purposes and meanings.[44] In contrast to branches, sequels follow the conclusion of a given phase of a major naval/joint operation. They are based on the possible outcomes of the preceding major operations. Sequels are the means of allowing the commander to exploit tactical or operational success.[45] A failure to plan sequels, whether because of lack of foresight or insufficient forces, has often had serious consequences for the courses and even outcomes of a major naval operation. Sequels are prepared in a rudimentary form—usually as an outline—in advance, so that subordinate tactical commanders can become fully familiar with them. The forces that would carry out a sequel should be identified and the sequel's objective clearly determined. Individual phases can be viewed as sequels to the basic plan.[46]

Operational Phasing

The ultimate objective is normally accomplished by spacing a major naval operation into several *phases*, related in time and space. In general, a phase is the time between the accomplishment of two successive intermediate objectives. Depending on the accomplishment of the intermediate objectives, phases can be planned to take place simultaneously or sequentially. Their main purpose is to stagger a major naval operation into several parts; otherwise, forces can approach or overshoot the point of culmination before the next intermediate objective is accomplished. The naval operational commander should not, however, arbitrarily break down a major naval operation into phases and thereby unnecessarily slow down the operational tempo.

Most offensive major naval operations will be conducted in a single phase while employing all available forces at high intensity. The exception comprises amphibious and anti-amphibious landing operations and major operations against enemy maritime trade or in defense and protection of friendly maritime trade. While a tactical action may consist of several phases, they usually overlap each other; hence, there is no clear break in fighting. Phases in a major naval operation are, in contrast, usually much longer and are marked by clear breaks in the intensity of fighting.

A change in the phases of a major naval operation usually involves changes of objectives/tasks. Normally, changes in objectives/tasks—not a timeline—should determine the phasing of a major naval/joint operation. Other factors include the distance from the main or intermediate base of operation, logistical resources, protection of logistical lines of supplies, terrain, and weather, to name a few. The individual phases gain significance only in the larger context of the operation.

The links between two consecutive phases of a major naval operation are most important for success. Planners should clearly define the conditions to be created for a transition from one phase to another. It is also critically important to hide from the enemy when that transition actually occurs. This can be achieved by concurrently intensifying actions in a false direction or by conducting deception.

Distinctions between consecutive phases should be made by separation in time or space, differences in the objectives to be accomplished, or variations in the forces assigned. Hence, each phase should aim at some intermediate objective necessary to the accomplishment of the ultimate objective. Phases are planned by identifying transition points between successive major tactical actions or by identifying changes in tempo within a particular combat action.

Anticipating the Point of Culmination

The naval operational commander and his planners should evaluate the elements of the operational idea that most directly affect the arrival at the culminating point. Specifically, ultimate and intermediate objectives and their sequencing, phasing, reserves, surprise, deception, and protection of the friendly center of gravity are closely related to the point of culmination. A culminating point can be prevented by determining correctly the number of intermediate objectives and sequencing them soundly. It can also be prevented by planning an operational pause after a given intermediate objective has been accomplished but before actions to accomplish the next intermediate objective are started.

Solid operational intelligence is critically important for planning a point of culmination. Reliable and continuous sustainment is another critical factor if the effects of culmination are to be overcome. The logisticians should accurately predict support and supply requirements in conjunction with the effects caused by combat losses, time-and-distance factors, lines of communications, weather and terrain, and repair and replacement capabilities. With the assistance of intelligence officers, logisticians can help target the enemy's critical logistics elements, which may cause an enemy offensive to reach a culminating point before reaching its objectives.[47]

Operational Pauses

In a major amphibious or anti-amphibious landing operation or in a major naval operation against enemy maritime trade and in defense and protection of friendly maritime trade, it is necessary to plan for relative breaks in fighting. *Operational pauses* should be part of a sound plan for a major naval operation. Time between consecutive phases would vary in length depending on losses and logistical sustainment capabilities. These pauses can be used to bring up new forces or regroup existing ones, replenish and refuel, and provide rest to personnel.

Properly understood, the term "pause" means that a major part of the main forces slackens or even drastically reduces its efforts while the supporting forces intensify their pressure on the enemy's main forces. A series of almost continuous smaller naval tactical actions will be conducted during an operational pause, to create thereby a favorable operational situation in the theater and prepare for the conduct of consecutive major naval operations.[48]

Pauses are highly useful for effective sequencing and synchronization of major naval tactical actions within a major naval operation. Their main purpose is to allow sufficient time to regenerate combat potential and avoid arrival at or overreaching the point of culmination. An operational pause can also be planned to shift friendly forces from one sector of effort to another. If friendly forces are on the defensive, the naval operational commander may order an operational pause to allow his forces to shift so they can operate along a more favorable line elsewhere.

The naval operational commander might intentionally and temporarily cede the initiative to the enemy so as to prepare a major naval operation in some other part of the theater or deceive the enemy about his real intentions. An operational pause should be planned by taking the anticipated point of culmination into account, so that the enemy cannot seriously threaten forces while they temporarily reduce the intensity of their actions.

Operational Tempo

Operational tempo refers to the pace at which a major naval operation moves toward the accomplishment of its ultimate objective. In general, an operation-sized force can move only as fast as its force elements' tactical rate of advance. The higher the operational tempo, the better the chance of maintaining the initiative. The initiative is largely a product of maintaining a high tempo, or rate of progress toward the completion of the assigned objective. It is also understood as the rate or rhythm of activity relative to the enemy's activity within a specific combat action. It includes the capability of forces to make a successful transition from one posture to another—that is, from offense to defense or vice versa.[49]

An essential factor in maintaining a fast operational tempo is *operational agility*—the ability of forces to act or react faster than the enemy. Agility depends on the speed of one's decision-making cycle. It can be enhanced by spatially dislocating the enemy's forces, through either maneuver or feints. A high operational tempo can be achieved by conducting combat actions simultaneously rather than in sequence. In addition, operational tempo is greatly increased by decentralized execution through mission command orders. Operational tempo further depends on the ability of a part of the force to operate at high intensity while the remainder of the forces regenerate combat power.

Operational Reserves

The operational idea should also address in broad terms the forces' size and mix and the location of the operational reserve. An *operational reserve* is a force of such size and combat potential as to have a decisive impact on the outcome of a major naval operation. An operational reserve is usually a one-time-use force.[50] It is used to ensure and consolidate operational success.[51]

The operational idea for a major amphibious landing might include a part of the landing force as a reserve after the first wave has landed. Such a reserve can be committed to battle ashore in case of unexpectedly strong enemy resistance or the appearance of other enemy forces in the area. It can also be used for conducting a feint or demonstration to distract the enemy's attention or tie down his forces along a certain part of the coast. The size of the reserve would vary, but it is generally much smaller than an operational reserve for a major land operation. In general, the potential impact on the outcome of the landing operation determines the force's size. In some Allied landings in World War II, reserves were the size of one reinforced division. Some elements were embarked on the ships in the proximity of the landing beaches, and some were on-call reserves, assembled at some distance from the landing objective area.

In the Allied invasion of Sicily in July 1943, both the British Eighth Army and the U.S. Seventh Army had what would be considered in land warfare tactical reserves. The U.S. Seventh Army's reserve consisted of four distinct and widely separated parts: the 2nd Armored Division (minus Combat Command A), reinforced by the 18th Regimental Combat Team (RCT) of the 1st Division and deployed with amphibious forces; the 82nd Airborne Division, on call after H-hour; the 39th RCT of the 9th Division, plus the 9th Division Artillery in North Africa, ready to move at any time after D-day; and the remainder of the 9th Division.[52]

Regeneration of Combat Potential

The operational design for a major naval operation should include ways and procedures to *regenerate forces' combat potential.* Phasing and operational pauses are primarily used to plan for sufficient time to regenerate combat power. After evaluating the enemy's strength, the naval operational commander can determine his *relative combat power* and calculate where his forces are in relationship to their culminating points. This information can be used in sequencing tactical actions, in pausing at the most opportune time, or, when on the defense, in designing actions that will lead to the opponent's culmination.[53]

Conclusion

The operational idea is the very heart of any operational design. The larger the military objective, the broader the operational perspective will be. Therefore, the scope of the operational idea for a campaign and that for a major operation are different in terms of time-space-force factors. The fewer elements an operational idea has, the less complicated it is and the more quickly it can be executed. However, none of the elements should be discarded without due analysis of the mission and the situation. The operational idea should be written in free form; shortcuts and buzzwords should be avoided. It should express in clear and commonly understood operational terms how the operational commander intends to sequence and synchronize all available sources of power to accomplish the assigned strategic or operational objective in the shortest possible time and with the fewest friendly losses.

The operational idea for a campaign is developed first, because the strategic objective always dominates the operational objectives. It also provides a framework for the operational idea of each subordinate major operation. Regardless of its purpose, an operational idea should be written concisely and clearly. At the same time it should ensure the accomplishment of the assigned military objectives. Ideally, it should surprise the enemy and impose one's will and initiative. The first employment of forces in a campaign is usually the most decisive. The naval operational commander and his planners should make judicious choices, from among all possible elements, as to which ones are to become part of the scheme.

Notes

1. Walter A. Vanderbeek, *The Decisive Point: The Key to Victory* (Fort Leavenworth, Kans.: School of Advanced Military Studies, U.S. Army Command and General Staff College, April 1988), p. 39.
2. Terence Robertson, *Channel Dash: The Fantastic Story of the German Battle Fleet's Escape through the English Channel in Broad Daylight* (New York: E. P. Dutton, 1958), pp. 52, 54–55, 175; Stephen W. Roskill, *The War at Sea 1939–1945*, vol. 2, *The Period of Balance* (London: Her Majesty's Stationery Office, 1956), p. 160.
3. Michael C. Vitale, "Jointness by Design, Not Accident," *Joint Force Quarterly* (Autumn 1995), p. 27.
4. Roy Appleman et al., *The War in the Pacific: Okinawa—The Last Battle,* United States Army in World War II, CMH Publication 5-11 (Washington, D.C.: Center of Military History, United States Army, 1948; repr. 1991), p. 473; some sources put U.S. casualties at 65,631 killed or wounded, while the Japanese lost seventy thousand troops and an equal number of civilians; Robert H. Scales, Jr., "Adaptable Enemies: Achieving Victory by Avoiding Defeat," *Joint Force Quarterly* (Autumn/Winter 1999–2000), p. 8.

5. Vanderbeek, *The Decisive Point*, pp. 34–35.

6. Gordon M. Wells, *The Center of Gravity Fad: Consequence of the Absence of an Overarching American Theory of War* (Arlington, Va.: Institute of Land Warfare, AUSA, March 2001), p. 4.

7. Malcolm W. Cagle and Frank A. Mason, *Sea War in Korea* (Annapolis, Md.: Naval Institute Press, 1957), p. 80.

8. Richard W. Bates et al., *The Battle of the Coral Sea, May 1 to May 11 Inclusive, 1942: Strategical and Tactical Analysis* (Newport, R.I.: Naval War College, 1947), app. II, p. v.

9. Phillip Kevin Giles and Thomas P. Galvin, *Center of Gravity: Determination, Analysis, and Application* (Carlisle Barracks, Pa.: Center for Strategic Leadership, U.S. Army War College, 31 January 1996), p. 19.

10. Myron J. Griswold, *Considerations in Identifying and Attacking the Enemy's Center of Gravity* (Fort Leavenworth, Kans.: School of Advanced Military Studies, U.S. Army Command and General Staff College, 14 May 1986), p. 28; Giles and Galvin, *Center of Gravity*, p. 19.

11. Elmer B. Potter and Chester W. Nimitz, eds., *Seemacht: Eine Seekriegsgeschichte von der Antike bis zur Gegenwart* (Hersching: Manfred Pawlak, rev. ed., 1986), originally published as *The Great Sea War: The Story of Naval Action in World War II* (Englewood Cliffs, N.J.: Prentice Hall, 1960), p. 187; "The Naval War in the Pacific," interrogation of Shigeru Fukudome, VADM, IJN, Interrogation NAV No. 115, 9–12 December 1945, USSBS (Pacific), Naval Analysis Division, *Interrogations of Japanese Officials* (Washington, D.C.: U.S. Government Printing Office, 1946), vol. 2, p. 504.

12. "SHO Operations and Taiwan Air Battle" (excerpts from Admiral Soemu Toyoda, *The End of the Imperial Navy* [Tokyo: April 1950]), folder G-14, box 7, record group 23: *World War II, Battle Evaluation Group (BEG)*, 1946–59, Naval Historical Collection, Naval War College, Newport, R.I., p. 8.

13. H. Engelmann, "Die Sicherstellung von Seeoperationen," *Militaerwesen* 3 (March 1980), p. 69.

14. Peter J. Palmer, *Operational Main Effort and Campaign Planning* (Fort Leavenworth, Kans.: School of Advanced Military Studies, U.S. Army Command and General Staff College, 1991), pp. 9–10.

15. Roskill, *The War at Sea 1939–1945*, vol. 2, pp. 94–95; Karl Doenitz, *Zehn Jahre und Zwanzig Tage* (Bonn: Athenaeum-Verlag, 1958), pp. 195–97.

16. Engelmann, "Die Sicherstellung von Seeoperationen," p. 69.

17. P. Navoytsev, "Zakonomernosti, Soderzhaniye i Kharakternye Cherty Sovremennykh Morskikh Operatsii" [Regularities, Content and Characteristics of Modern Naval Operations], *Morskoy Sbornik*, no. 7 (July 1986), p. 21.

18. John W. Foss, "Command," *Military Review* 5 (May 1990), p. 4; Palmer, *Operational Main Effort and Campaign Planning*, p. 8.

19. Friedrich von Bernhardi, *On War of To-Day*, vol. 2, *Combat and Conduct of War*, trans. Karl Donat (London: Hugh Rees, 1913), pp. 361–62.

20. R. Ernest Dupuy, Trevor N. Dupuy, and Paul F. Braim, *Military Heritage of America* (New York: McGraw-Hill, 1956), vol. 1, p. 15.

21. Vojno-Pomorski Skolski Centar, *Zajednicke Operacije Vidova* (Split, Yugoslavia: Skripta Iz Operatike RM, V Lekcija [Projekt], 1970), pp. 44–45.

22. Michael Handel, "Intelligence and Deception," *Journal of Strategic Studies* (March 1982), p. 126.

23. Michael Handel, *War, Strategy and Intelligence* (London: Frank Cass, 1989), pp. 314–16.

24. Ibid., p. 314.

25. Charles Cruickshank, *Deception in World War II* (Oxford, U.K.: Oxford Univ. Press, 1979), p. 51.

26. Albert N. Garland and Howard McGaw Smyth, assisted by Martin Blumenson, *United States Army in World War II: The Mediterranean Theater of Operations—Sicily and the Surrender of Italy* (Washington, D.C.: Office of the Chief of Military History, Department of the Army, 1965), p. 65.

27. Cruickshank, *Deception in World War II*, pp. 51–52.

28. Commander in Chief Mediterranean, memorandum, "Operation Fracture," 27 June 1943, Strategic Plans Division Records, Op-607 Support Plans Branch [hereafter

Op-607], box 543, Operational Archives, Naval Historical Center, Washington, D.C. [hereafter NHC].

29. "Military Cover and Deception during World War II," Strategic Plans Division Records, Op-607 (series XXII), box 531, NHC, pp. I-11 and I-12.

30. Garland and Smyth, *The United States Army in World War II,* p. 65.

31. Cruickshank, *Deception in World War II,* p. 60.

32. Supreme Headquarters, Allied Expeditionary Force, "Plan Fortitude South II, 19 July 1944," Strategic Plans Division Records, Op-607 (series XXII), box 531, NHC, p. 13.

33. Cover and Deception Report ETO, "Cover & Deception, Synopsis of History," Strategic Plans Division Records, Op-607 (series XXII), box 549, NHC, p. 9.

34. John D. Johnson, *Sequencing Operations: Considerations for the Operational Planner* (Fort Leavenworth, Kans.: School of Advanced Military Studies, U.S. Army Command and General Staff College, 20 May 1991), p. 4.

35. Richard J. Dixon, *Operational Sequencing: The Tension between Simultaneous and Sequential Operations* (Fort Leavenworth, Kans.: School of Advanced Military Studies, U.S. Army Command and General Staff College, 1994), p. 5.

36. Bates et al., *The Battle of the Coral Sea, May 1 to May 11 Inclusive, 1942,* p. 15.

37. John B. Lundstrom, *The First South Pacific Campaign: Pacific Fleet Strategy, December 1941–June 1942* (Annapolis, Md.: Naval Institute Press, 1976), pp. 98–99.

38. Dixon, *Operational Sequencing,* p. 34.

39. S. Filonov, "Morskaya Operatsiya" [Naval Operation], *Morskoy Sbornik* 10 (October 1977), p. 25.

40. Hans-Martin Ottmer, *"Weseruebung": Der deutsche Angriff auf Daenemark und Norwegen in April 1940* (Munich: R. Oldenbourg, 1994), pp. 37–41, 52–56, 79–84; Walter Hubatsch, *"Weseruebung": Die deutsche Besetzung von Daenemark und Norwegen 1940. Nach amtlichen Unterlagen dargestellt mit einem Anhang: Dokumente zum Norwegenfeldzug 1940,* 2nd rev. ed. (Goettingen: Musterschmidt Verlag, 1960), pp. 42, 49.

41. Douglas MacArthur, *Historical Report on Allied Operations in Southwest Pacific Area,* vol. 2, *Japanese Operations in the Southwest Pacific Areas, 8 December 1941 to 2 September 1945* (Washington, D.C.: U.S. Government Printing Office, 1966), chaps. 8–14, p. 340ff.

42. Richard W. Bates et al., *The Battle for Leyte Gulf, October 1944: Strategical and Tactical Analysis,* vol. 1, *Preliminary Operations until 0719 October 17th, 1944 Including Battle off Formosa* (Newport, R.I.: Naval War College, 1953), pp. 459–60.

43. William S. Pennypacker, *Sequels: Thinking about the Future* (Fort Leavenworth, Kans.: School of Advanced Military Studies, U.S. Army Command and General Staff College, May 1988), p. 34.

44. A sequel is defined by the *American Heritage Dictionary* as "anything that follows or is in continuation; it also refers to a result or consequence."

45. Pennypacker, *Sequels,* p. 4.

46. L. D. Holder, "A New Day for Operational Art," *Army* 3 (March 1985), pp. 4–5.

47. David J. Benjamin, *Prerequisite for Victory: The Discovery of the Culminating Point* (Fort Leavenworth, Kans.: School of Advanced Military Studies, U.S. Army Command and General Staff College, 13 May 1986), pp. 26–27.

48. Ibid.

49. Johnson, *Sequencing Operations,* pp. 7–9.

50. Albert F. Turner, Jr., *The Operational Reserve: What Should It Be Used For?* (Fort Leavenworth, Kans.: School of Advanced Military Studies, U.S. Army Command and General Staff College, 6 April 1988), p. 22.

51. Terry W. Bullington, *Considerations for the Organization and Employment of an Operational Reserve* (Fort Leavenworth, Kans.: School of Advanced Military Studies, U.S. Army Command and General Staff College, 16 May 1986), p. 3.

52. Garland and Smyth, *The United States Army in World War II,* pp. 97–98.

53. David M. Cowan, *The Utility of the Operational Pause in Sequencing Battles to Achieve an Operational Advantage* (Fort Leavenworth, Kans.: School of Advanced Military Studies, U.S. Army Command and General Staff College, 1987), p. 24.

CHAPTER SIX

Preparation and Execution

When the planning process is completed and the basic and supporting plans have been written, the next phase is that of preparation. The more time available for preparation, the higher the likelihood of the successful execution of the plan.

Preparation

The preparation of a major naval operation is far more complex and time consuming than that for a naval tactical action. It encompasses a comprehensive system of measures and actions, of which one of the most important is the organization of comprehensive combat support. These measures should be carried out in the shortest possible time and, if possible, concealed from the enemy. The naval operational commander needs to obtain accurate, relevant, and current information about the locations and movements of the enemy forces. In the preparatory phase of a major naval operation one of the primary tasks is to collect, evaluate, and analyze data on all the aspects of the operational situation, especially the intentions and actions of the enemy forces to be deployed in the sector of main effort. Another major task is the proper organization and timely conduct of rehearsals for forces taking part. The entire operation and its segments should be thoroughly rehearsed in a series of planning games and/or war games. Changes should be made in the plan based on the lessons learned from these games. In addition, various parts of the operation plan should be tested by individual force elements at sea.

Some preparatory measures can be completed before the start of a major naval operation. For example, surveillance and reconnaissance of the part of a maritime theater in which the operation will take place should be conducted continuously but well before naval forces are deployed. Offensive minefields can be laid covertly by using innocuous commercial or fishing vessels as a part of preparations for the initial major naval operation. Likewise, defensive minefields can be laid and prepared for activation before the outbreak of the hostilities.

Execution

The execution of a major naval operation encompasses several interrelated and often overlapping segments: deployment, actions of forces in the main and secondary sectors, accomplishment of the operational objective and consolidation of operational success, and withdrawal to home base or operational redeployment to another part of the maritime theater.

Most major naval operations conducted today would be relatively short and encompass a single phase. However, they would be conducted at much higher intensity than in the past, and all available forces would probably be employed. Phases of a major naval operation in execution invariably differ from the initial basic plan, both in the time of occurrence and in duration. Phases might change because the enemy offers stronger or weaker resistance, intermediate objectives change, additional forces become available, or the sector of main effort needs to be changed. A series of almost continuous smaller tactical actions will be conducted during operational pauses to create a favorable operational situation in the theater and to prepare for the conduct of consecutive major naval operations.

The chances of successful execution of a major naval operation are greatly increased by high combat readiness of friendly forces, timely and covert deployment, short tactical maneuver of friendly forces to their respective areas of operation, and continuous reconnaissance and surveillance of the enemy naval forces in peacetime and as the operation is executed. Other prerequisites for success include focusing efforts against the most important part of the enemy forces, executing timely supporting strikes, providing reliable cover from air strikes for the main forces at sea and the most important naval bases/airfields, and creating a situation favorable for the forces operating in the sector of main effort.

The most decisive factor for the success of the operation as a whole is destruction or neutralization of enemy C4 systems. The uninterrupted work of one's communications and electronic sensors are critical for the success of the entire operation. However, most important of all will be close cooperation among naval combat arms and combat arms of other services.

Deployment

Deployment is the key prerequisite for the success of any plan for a major naval operation. In deploying naval forces, one cannot easily correct errors in the forces' size, mix, and timing once hostilities at sea start, if it can be done at all. The outcomes of initial clashes at sea and in the air will be critical in the overall success of a major naval operation, because they will largely determine the conditions for successive combat actions.

The essence of the deployment is the timely phasing-in of friendly forces to their assigned combat areas. Naval forces, in contrast to ground forces, are deployed in areas possibly under enemy control. Deployment can be conducted partially drawing upon forces already at sea. Other forces taking part in a major naval operation would sortie out from several bases, spread sometimes over several hundred nautical miles. Their operating area might be many hundreds or even thousands of nautical miles from the scene of action. If forces are forwardly deployed, as are U.S. Navy forces currently in the western Pacific and in the Persian (Arabian) Gulf, in the case of hostilities they would conduct only an operational or even merely tactical deployment. In contrast, deployment of naval forces for a major naval operation in the littorals would be predominantly tactical in size because of the generally much shorter distances and cluttered environment.

Deployment of one's naval forces is normally conducted from the base of operations to the area of concentration. The initial base of operations dictates the position, directional orientation, and length of lines of operations for friendly naval forces. The direct route between a given base and the assigned physical objective is the shortest and the best. However, often a circuitous route must be used to avoid a strong enemy position or enhance the element of surprise. This is especially true in a narrow sea endowed with a large number of offshore islands or archipelagoes. The essence of the deployment of a given force element is the speed of movement, and the time required to transit the given distance. On the open ocean, naval forces would sometimes carry out their deployment in the waters either under the enemy control or where control is in dispute. In contrast, a small navy will deploy within its own coastal waters or sometimes to sea areas under the enemy control. Also, forces taking part in a major naval operation in an enclosed or semi-enclosed sea can be deployed from their bases simultaneously or sequentially.

Tactical Maneuver

Tactical maneuver is a planned movement of individual force elements out of their basing areas toward assigned sectors/zones in which the main actions would take place.[1] The sequence of movement of force elements should be based on the speeds of advance and transit routes. The naval operational commander and his staff have to determine the optimal cruising formation offering the best possibility for using weapons and mutual support in defense against a possible enemy attack. Forces participating in a major naval operation should conduct their movements out in a timely fashion to their respective sectors of main and secondary effort.

During the execution of the tactical maneuver, forces should have sufficient degrees of control of the basing/deployment area. In an enclosed sea theater, tactical maneuver should be conducted during the night. Optimally, the entire operation should be completed during the time of darkness. To ensure survivability of friendly forces during the execution of tactical maneuver, it is necessary to alert coastal defenses and maintain them at the highest state of combat readiness. In addition, special groups for securing flanks of friendly forces during their maneuver might be organized.

The key factors for the timely tactical maneuver of friendly forces are continuous and reliable reconnaissance and surveillance of the enemy forces, selection of the proper routes of movement and correct calculation of the speed of movement, and determination of sufficient reserve time for the maneuver of friendly forces. Other elements include overcoming interference by the enemy forces, maintaining high readiness of forces, and the timely concentration of forces in advanced bases and airfields.[2] Tactical maneuver in enclosed sea theaters should be carried out covertly and within a short time frame. Covertness of movement of friendly forces can be greatly enhanced by precluding enemy reconnaissance and surveillance of the friendly basing/deployment area, by making use of night and bad weather conditions, by shifting routes of movement for friendly forces to the area where the enemy would have greater difficulties in detecting and monitoring their movements, and by conducting feints or full-scale operational deception.

Tactical maneuver of friendly forces should be supported by a series of diverse actions against possible interference by the enemy forces. Such actions include strikes against enemy ASW forces, suppression or neutralization of enemy air defenses, destruction of enemy surveillance and reconnaissance forces, and disorganization of the enemy command and control.[3]

Actions in the Sectors of Main and Secondary Effort

The most decisive phase of any major naval operation is conducting actions in the sector of main effort. Regardless of the operation's main purpose, the major part of forces in a major naval operation will be employed in the sector of main effort. The most critical element for the ultimate success of the entire operation will be to obtain and maintain the initiative. The first missile strikes should be massed and conducted on a sustained basis. At the beginning of the operation, the sector of the main effort should be kept secret from the enemy, and feints should be conducted in false directions.[4] Strikes would be carried out against the enemy's main forces operating in the main and secondary sectors. The purpose will be to create favorable conditions to resolve the main task of the entire operation.[5] Friendly forces assigned to the sector of main effort should be those with the highest combat potential because they must be able to destroy the main enemy forces. This will be accomplished through timely spreading out and by carrying out simultaneous or sequential missile strikes.[6]

For the naval operational commander and his staff one of the most complex problems to resolve is to determine the proper targets for missile or torpedo strikes, and the area and the time of actions of the main forces. This is an especially critical factor in an enclosed sea theater, when the main forces will be predominantly composed of small surface combatants. The use of mines can either facilitate or make more difficult the employment of friendly forces in the pending operation. Properly laid mines can limit the freedom of movement for the enemy forces. At the same time, the faulty use of mines can considerably limit the movement of naval forces in the course of a major naval operation.

In a fleet-versus-fleet engagement or in an attack against enemy amphibious forces or convoys on the open sea, missile or torpedo strikes against the main enemy forces should be carried out simultaneously, to inflict the greatest degree of destruction.[7] However, if conducted at the wrong time and place or against the wrong targets, such strikes can considerably undermine the very purpose of the operation. Strikes should result in the destruction of the enemy's forces and quickly achieve the operation's objective. When the defender has only limited forces this would require the naval operational commander and his staff to determine properly the ratio of forces for the main strike and those used for developing success, not only in regard to their sizes but also compositions, so as to achieve required results and effects in a given phase of a major naval operation.

In a surface-to-surface engagement, the most potent weapons are antiship missiles and torpedoes fired by submarines or surface ships, combined with strikes by fighter-bombers and attack aircraft. In an engagement in littoral waters, coastal antiship missile/gun batteries can take part. Hence, it is of prime importance that a decision to conduct strikes be based on the most accurate and timely information on the operational situation in a given part of the theater. This, in turn, requires continuous reconnaissance and surveillance of the basing and operating areas of the enemy forces.

To achieve success in the main sector, it is necessary to neutralize enemy covering, supporting, or patrol forces or to prevent them from interfering with the actions of friendly forces in the main sector of effort.[8] In that case, a part of the friendly forces has to conduct actions in the secondary sector of effort. These actions usually have the character of preliminary support. They are conducted concurrently with those in the sector of main effort. The main purpose of supporting actions is to neutralize enemy ASW in a given part of the theater, weaken or neutralize the enemy's air defenses, and destroy or neutralize his C4 nodes.[9] The most decisive contribution to success would be destruction or neutralization of the enemy C4 nodes for the time when an operation takes place.

Operational Success

The main purpose of a major naval operation is to accomplish operational success as quickly as possible, by destroying or neutralizing the major part of the enemy's main forces. In some cases, operational success can be attained by the action of friendly supporting forces.

Experience shows that operational success can be achieved even if the operational objective has been improperly chosen. For example, the Japanese, in their surprise attack on Pearl Harbor on 7 December 1941, inflicted heavy losses (five battleships sunk and three damaged, three light cruisers damaged, and three destroyers and several other ships sunk) but not a mortal blow on the U.S. Pacific Fleet, since by coincidence two U.S. fast aircraft carriers plus seven cruisers, fourteen destroyers, and four submarines were at sea. In addition, two heavy cruisers, three light cruisers, twenty-four destroyers, four destroyer-minelayers, eight fast minesweepers, five submarines, and numerous auxiliary ships were undamaged.[10] Nevertheless, the Japanese achieved an operational success that allowed them to complete the conquest of the Philippines without interference from the U.S. Pacific Fleet.

Sometimes the attacker came close to operational success but, either through poor intelligence or sheer lack of will, fails to grasp the opportunity. An example of operational success in the battle for Leyte was the Allied victory over the Japanese forces in the Surigao Strait—it prevented the Japanese from attacking Allied troops and supplies ashore in Leyte Gulf. The Japanese missed a chance to accomplish operational success because Admiral Kurita failed to exploit his tactical success in the battle off Samar.

Once operational success is attained, it must be followed quickly with actions aimed to exploit that success. If the enemy is not decisively beaten, he might endanger the success achieved. In the case of a major offensive naval operation, the operational success must be consolidated by pursuing and destroying remaining enemy forces. The shift from the major combat phase to pursuit should be seamless. However, in some cases an operational pause after the pursuit might be required to regenerate combat potential. After the pursuit, forces might withdraw to their base or be redeployed for other tasks in the same or an adjacent theater. For example, the U.S. Navy achieved a great operational success in the battle of Midway on 3–6 June 1942. However, that in itself would not have been sufficient to reverse the tide of war in the Pacific had it not been followed by an offensive operation to consolidate the success achieved at Midway. Thus, after some debate between the U.S. Army and Navy, on 2 July 1942 the invasion of Guadalcanal began, and with it started what later emerged as the Solomons campaign.[11]

Return to Home Bases/Redeployment

After the ultimate objective of a major naval operation is accomplished and operational success is consolidated, friendly forces usually return to their home bases or are redeployed to another area of operations, in the same or an adjacent maritime theater, to obtain a new, more advantageous base of operations for a subsequent operation or to exploit operational success. The need to obtain and maintain the initiative in the war at sea requires continuous and decisive execution of combat action without long pauses between operations. In determining basing ports for returning forces the naval operational commander must take into consideration not only restoration of their combat capability but also the need to use bases nearby, so that forces can be spread out again from them for the next operation.

Command and Control

The main responsibilities of the naval operational commander in the course of a major naval operation include conducting a running estimate of the situation; supervising and influencing subordinate tactical commanders' actions; changing intermediate objectives; changing command relationships and subordination; consolidating operational or strategic success; and preparing for the next phase of a major naval/joint operation.

Once the initial situation is assessed, the decision made, and plans and orders complete and executed, the naval operational commander and his staff should continue monitoring the situation and conduct a *running estimate of the situation*. It is critical to assess the situation quickly and then make a new decision, which will in turn require modifications or changes in new plans or orders. This is critically important, because modern combat is highly dynamic and any estimate of the situation will undergo constant and sometimes drastic changes.[12]

Besides conducting a running estimate of the situation, the naval operational commander influences his subordinate commanders' accomplishment of the mission by changing their objectives and tasks, reallocating or reassigning combat forces and support forces from one sector or area to another, reallocating and reassigning combat support, changing subordinate commanders' responsibilities, or, if necessary, relieving subordinate commanders.

The naval operational commander's intent should provide his operational vision and thereby enable his subordinate tactical commanders to understand clearly what actions they need to take in the accomplishment of the commander's overall objective. The naval operational commander's intent should ensure sufficient freedom of action for the subordinate tactical commanders in the accomplishment of their assigned missions; otherwise, unity of effort cannot be achieved. Likewise, in carrying out their assigned tasks, subordinate service/functional component commanders should keep in mind the naval operational commander's intent. In exercising initiative, subordinate tactical commanders should act in consonance with that intent. The naval operational commander's freedom of action is greatly enhanced through decentralized decision making, based on a clear statement of intent and use of mission command orders. In many cases, the operational commander will be far removed from subordinate tactical commanders. The greater the separation, the greater the danger that his decisions will lose force and that his intent will not be passed down through the lower levels of the chain of command.

The naval operational commander should be mentally ready to modify, change, or even abandon his plan in response to unexpected events after a major operation starts. He should retain operational flexibility but at the same time not lose sight of the main objective. The advantages of a situation can never be fully utilized if subordinate commanders wait for orders. The employment of large forces can be successful only if commanders at all levels are competent and accustomed to independent action—hence the need in time of peace to instill in subordinate commanders the habit of acting independently. Only then can the naval operational commander hope that subordinates will carry out his intent. When in doubt, it is always preferable to act, and thereby obtain and maintain the initiative, rather than to react to the enemy's actions.[13]

Optimally, the naval operational commander should *supervise his subordinate tactical commanders* in such a way as to encourage initiative and independent thinking.[14] He should interfere with the decisions of his subordinate commanders *only* when those decisions are unsound and could endanger the outcome of the entire mission. The naval operational commander should have great confidence in the skills and abilities of his subordinate operational-tactical and tactical commanders, since he is usually removed by hundreds, or even thousands, of miles from the scene of combat. The greater the distance, the higher the possibility that his decisions will reach subordinate commanders too late or that those decisions will not correspond to a fast-moving situation. Therefore, he should act with speed and energy to make sure that his intent and orders are clearly understood and carried out.

Under some circumstances, a subordinate commander should modify or even abandon the assigned mission if such an action is necessary to satisfy the operational commander's intent. The subordinate commander should seek approval first, but if that is not possible, he should take full responsibility for his decision. Afterward, a subordinate tactical commander should explain and justify the action to the operational commander's satisfaction.[15] Subordinates should also make every effort to maintain contact with their operational commander. They should keep him fully informed of any significant change in the situation, while exhibiting initiative and energy in resolving problems within their areas of responsibility. They should be encouraged to propose changes in orders and directives based on their own assessments of the situation.[16]

Normally, the plan of a major naval operation cannot be successful without a series of well-integrated actions and measures to *shape the battlefield,* a process conducted before and during the combat. This is accomplished principally by destroying or degrading the enemy's critical capabilities; delaying or preventing the deployment or movement of enemy forces into the theater; manipulating the enemy's perceptions; degrading or neutralizing the enemy's ability to make decisions; and obtaining and maintaining one's freedom of action. The destruction or degradation of the enemy's critical capabilities can be accomplished by attacking a wide range of targets in the theater, such as C4 nodes, elements of logistical support and sustainment, tactical air support, and air defenses. C2W is one of the most important and potentially most effective ways to shape the theater, by destroying or neutralizing the enemy's ability to obtain an accurate picture of the situation and make sound and timely decisions. The naval operational commander should use both lethal and nonlethal *operational fires* as appropriate to isolate the area of operations in which a new major naval operation will start. Lethal operational fires remain, for the foreseeable future, the most effective way of preventing enemy forces and follow-on forces from arriving in the theater, of interdicting or completely cutting off enemy forces from their sources of supply and sustainment, and in some cases of cutting off the retreat of the pursued enemy forces beyond the effective range of ground forces.

The naval operational commander should pursue resolutely the ultimate objective of a major naval operation. He should not be distracted by some relatively insignificant event and shift or even abandon his original objective. In a dynamic combat situation, however, rigid adherence to the original objective can lead to a major setback, or even the failure of the entire major naval operation. The operational situation might change because drastic changes in any one of the factors of space, time, or force require reestablishing balance or scaling down the objective. In some cases, unexpectedly weak enemy resistance can lead to a *reassessment of the original objective,* resulting in the selection of some other, larger objective.

Once the enemy operational center of gravity is determined, it should serve as the focus for friendly courses of action during the naval operational commander's estimate of the situation and the heart of his basic plan for the operation. The enemy center of gravity should be reassessed constantly throughout the major naval operation.[17] This evaluation should concentrate on significant changes in the strategic or operational situation. Any change in objective should lead to a reevaluation of the enemy's critical strengths and weaknesses and to redetermination of the enemy's center of gravity.

The operational commander should not become a slave to the plan; yet he should not alter a plan with which all are familiar unless a changing situation creates problems or opportunities that cannot be overlooked. The naval operational commander should accept the inevitability of uncertainty in the situation and take advantage of it before the enemy commander does. When the situation demands, the naval operational commander may use a quick estimate of the situation for decision making in the minute-to-minute, hour-to-hour conduct of operations. The art of operational leadership consists of timely recognition of circumstances and of the moment when a new decision is required. The U.S. general Albert C. Wedemeyer (1897–1989) argued that "better a faulty plan or decision permeated with boldness, daring, and decisiveness, than a perfect plan enmeshed in uncertainty."[18]

The successful use of *operational* deception can manipulate the enemy commander's perceptions. The ultimate purpose of these efforts should be to deceive the enemy as to one's intentions, capabilities, or both. Cover and deception of one's own forces and their actions have great importance for success. Hence, operational deception should be aimed at preserving the secrecy of the objective and the character of the forthcoming major naval operation. It should be designed to deceive the enemy so as to enhance the effectiveness of forces and keep losses to the minimum. This involves continuous action to destroy or neutralize enemy reconnaissance forces.[19]

The naval operational commander should respond to unanticipated threats or opportunities by shifting forces from less threatened sectors to new sectors of effort. This can occur because intermediate objectives or their priorities change. A shift in the main sector of effort can result from determining an entirely new objective. The naval operational commander should carefully reevaluate the situation in the theater and avoid being fixated on some spectacular enemy tactical action, thereby missing larger trends in the operational situation.

The naval operational commander should try whenever possible to impose his tempo on the enemy.[20] In visualizing a sequence of actions by friendly forces, the naval operational commander should expect that conditions in combat will change; he should then adjust to these changes by developing branches that address unlikely but possible enemy reactions. Likewise, plans should include sequels, ranked by priority, based on the probability of occurrence. In the course of a major naval operation, the operational commander should carefully and continuously monitor the actions in progress. He should also try to envision the situation and possible changes several weeks or even months ahead. In sequencing the actions of friendly forces, the naval operational commander should not foreclose future options. This is primarily a matter of building options into the initial operation plan, but it can also result from running estimates of the situation. The operational commander should anticipate and orchestrate the actions that logically follow each other.

In general, one of the naval operational commander's most difficult tasks is to predict and identify the culminating point and whether it has been exceeded, or reached. The naval operational commander should timely sense or anticipate the approach to, or arrival at, the point of culmination during a major naval operation, so that the ultimate objective can be accomplished. The naval operational commander who improperly balances ends and means would cause a mismatch between combat and sustaining resources. This, in turn, may bring the major naval operation to culmination before the ultimate objective is reached.[21]

A culmination point for one's naval forces is extremely difficult to anticipate in practice, because so many factors affect it. To complicate the matter, culmination at the operational level, in contrast to the tactical level, becomes apparent only some time after it has taken place. So it can be discerned only in retrospect. The naval operational commander's ability to prevent culmination of friendly forces while causing the enemy to reach his is one of the keys to operational success. This can be accomplished only by clear and precise knowledge of friendly and enemy combat power at all times—a hard thing to obtain.

The naval operational commander who is fixated on the current or next immediate combat action would have great difficulty recognizing the culmination, or combat power, of friendly forces in time. Therefore, he should think broadly and far beyond current actions, visualizing the decisions and actions that will be required to obtain, or regain and then maintain, the initiative. In such a way, he can anticipate strain and stress on his forces.

The naval operational commander may sometimes have more combat forces than logistic resources can support simultaneously. This is especially true in major joint/combined operations, when demands for logistic resources may be highly competitive. *Operational logistics and sustainment* are the responsibility of the theater commander. He applies logistic resources to generate, produce, and support theater combat power. Moreover, he should ensure that his plan integrates operation and logistics plans. While a major naval operation is in progress, the naval operational commander should continuously balance current consumption of materiel with the need to build up support for subsequent actions. Carefully planned and executed forward staging will allow the naval operational commander to maintain the desired operational tempo. On the other hand, failure to plan for logistical sustainment will invite premature culmination.

Overwhelming power can be generated and applied if the operation plan contemplates the establishment and buildup of bases, favorable lines of operations and communications, and the selection of decisive points on which to focus forces so as to cause the enemy to culminate first.[22] In his overall concept, the naval operational commander should establish a sound relationship between operations and logistics. He does this by insisting on close cooperation and exchange of information between operations and logistic planners. In developing the concept of logistic support, the naval operational commander should decide on the proper balance between centralization and decentralization of logistic functions and responsibilities. Too much centralization can result in sluggish or rigid response. The naval operational commander is concerned not only with the logistic support of the ongoing major naval operation but also with that of the successive operation. He needs a clear and comprehensive assessment of logistic support in order to design or modify his idea for the operation. The naval operational commander's staff should assess how logistic support issues will affect the sequencing of operations. They should also recommend options to compensate for logistic shortfalls.

Conclusion

Preparation and execution of a major naval operation is the culmination of the entire planning process. Without thorough planning, it is difficult to prepare and execute a major naval operation properly. An operation plan is rarely, if ever, executed as originally envisaged, because the enemy does not act the way one thinks he will. Friction and the fog of war will make actions on both sides highly unpredictable. Hence, only a highly flexible plan and mental agility on the part of the naval operational commander can ensure ultimate success. The operational commander should quickly adjust objectives and the methods for accomplishing them to changes in the situation. Once the operation starts, the naval operational commander must closely monitor changes in the situation, conduct quick running estimates, and make appropriate decisions. The key for success in the execution of a major naval operation lies in uninterrupted reconnaissance/surveillance and good intelligence on the enemy's situation, a high state of combat readiness and training of friendly forces, and a command and control process that ensures centralization in planning but allows room for subordinate tactical commanders to exercise initiative in the execution of their assigned tasks.

Notes

1. Gerhard Regner, ed., *Operative Kunst der Seestreitkraefte,* Teil I/2, trans. Heinz Engelmann et al. (Dresden: Friedrich Engels Military Academy, 1981), p. 111.

2. Ibid., pp. 112–13.

3. Ibid., p. 114.

4. Heinz Engelmann, "Die Sicherstellung von Seeoperationen," *Militaerwesen* 3 (March 1980), p. 70.

5. P. Navoytsev, "Zakonomernosti, Soderzhaniye i Kharakternye Cherty Sovremennykh Morskikh Operatsii" [Regularities, Content and Characteristics of Modern Naval Operations], *Morskoy Sbornik,* no. 7 (July 1986), p. 21.

6. Regner, ed., *Operative Kunst der Seestreitkraefte,* p. 128.

7. Adolf Feil, ed., *Grundlagen der Theorie der operativen kunst der SSK,* trans. Dieter Foerster and Bernd Kulbe (Dresden: Friedrich Engels Military Academy, March 1989; translation of *Operativnoye iskusstvo voyenno-morskovo flota*), p. 41.

8. Engelmann, "Die Sicherstellung von Seeoperationen," p. 69.

9. Navoytsev, "Zakonomernosti, Soderzhaniye i Kharakternye Cherty Sovremennykh Morskikh Operatsii," p. 21.

10. Juergen Rohwer and Gerhard Huemmelchen, *Chronology of the War at Sea 1939–1945,* 2nd ed. (Annapolis, Md.: Naval Institute Press, 1992), p. 104.

11. John Miller, *United States Army in World War II: The War in the Pacific—Guadalcanal: The First Offensive,* CMH Pub 5-3 (Washington, D.C.: Center of Military History, 1949; repr. 1989), pp. 16–17.

12. Eike Middeldorf, *Fuehrung und Gefecht. Grundriss der Taktik* (Frankfurt am Main: Bernard & Graefe Verlag für Wehrwesen, 1968), p. 59.

13. Daniel J. Hughes, *Moltke on the Art of War: Selected Writings* (Novato, Calif.: Presidio, 1993), pp. 132–33.

14. Cited in John T. Nelsen, *Where to Go from Here? Considerations for the Formal Adoption of Auftragstaktik by the US Army* (Fort Leavenworth, Kans.: School of Advanced Military Studies, U.S. Army Command and General Staff College, December 1986), pp. 10–11.

15. Ibid., p. 10.

16. Ibid., p. 11.

17. Phillip Kevin Giles and Thomas P. Galvin, *Center of Gravity: Determination, Analysis, and Application* (Carlisle Barracks, Pa.: Center for Strategic Leadership, U.S. Army War College, 31 January 1996), p. 18.

18. Bruce Condell and David T. Zabecki, eds., *On the German Art of War: Truppenfuehrung* (Boulder, Colo.: Lynne Rienner, 2001), p. 23.

19. Engelmann, "Die Sicherstellung von Seeoperationen," p. 70.

20. British Army, *Operations,* prepublication ed. (1989), pp. 3-14 and 3-15.

21. Henry S. Scharpenberg, *U.S. Armed Forces and the Operational Level of War: Are We Prepared to Win?* (Fort Leavenworth, Kans.: School of Advanced Military Studies, U.S. Army and General Staff College, May 1998), p. 5; David M. Cowan, *The Utility of the Operational Pause in Sequencing Battles to Achieve an Operational Advantage* (Fort Leavenworth, Kans.: School of Advanced Military Studies, U.S. Army Command and General Staff College, 1987), p. 23.

22. Robert T. Dail, *Does the U.S. Army Really Understand Operational War? A Logistics Perspective* (Fort Leavenworth, Kans.: School of Advanced Military Studies, U.S. Army Command and General Staff College, 2 May 1988), p. 33.

The Future

A major naval operation is the principal method of accomplishing operational objectives in a maritime theater. It can be the most effective way to achieve decisive results within a given time frame, bringing about a drastic change in the situation in a specific theater of operations. The only other way of achieving the same results is attrition warfare, which should be avoided even when the operational commander enjoys quantitative superiority. Major naval operations are an area of study of operational art that Western naval theoreticians and planners have generally neglected. Too much emphasis is given instead to the tactical employment of naval forces and to the employment of naval forces in operations short of war. The real danger is of creating the impression that major naval operations will not be conducted in the future; this could not be more wrong. The absence of any serious threat at sea today should not delude anyone into believing that naval forces will never be required to plan and conduct major operations in the future. In addition, any regional conflict may require that the U.S. Navy plan for and execute a major naval operation. Threats to national interests at sea tend to come without much warning. Fleets are built and maintained primarily not to conduct low-intensity conflict but to wage war, whether regional or global.

The types of major naval operations depend on the main purpose, the predominant sea or ocean areas in which the operation is conducted, the timing, and the degree of participation by other services. While naval forces will always play the most critical role in the accomplishment of operational or strategic objectives, other services and their combat arms also have roles to play. This will be especially true in littoral waters, because the proximity and influence of a landmass will allow not only land-based aircraft but also ground forces to take part in major naval operations. Therefore, not only amphibious landings but also major naval operations against enemy maritime trade or for the protection of friendly maritime trade will represent an increasingly joint or even combined activity.

The likelihood of conducting a major naval operation against the enemy fleet at sea is remote today. The most likely scenario will be a major naval operation aimed at destroying or neutralizing a major part of the enemy fleet at its bases and at sea. In an enclosed or semi-enclosed sea, such as the Persian (Arabian) Gulf, such an operation will also involve land-based air and in some cases also ground forces. Amphibious landings with operational objectives will be conducted in the future, but they will be much smaller in size than they were in the past. The blue-water navies, such as the U.S. Navy, will occasionally conduct major operations versus enemy shores aimed to destroy or neutralize important installations/facilities in the littorals and even in the strategic depth of the enemy's defenses. Major operations to attack enemy maritime trade and protect friendly maritime trade will be conducted mostly by joint employment of naval forces and other services. Most of the actions in support of friendly ground forces operating in the coastal area will be tactical in scale; major naval operations will be conducted only in exceptional situations.

In the course of the execution of a major naval operation, the operational commander should continuously monitor the situation and adjust his plans and orders accordingly. Frequent changes of plans should be avoided. More important, the principle of unity of command should be fully observed. Simplicity requires one to avoid the complex, the complicated, and the intricate; the plan should proceed in the most direct and natural manner toward the attainment of the assigned operational or strategic objective. Other factors being equal, the simplest plan executed promptly is preferable to a complex plan executed later.

The focus of theory should be on the type of major naval operation most likely to be conducted at present or in the near future. However, this does not mean the other types of major naval operations should be neglected or ignored. For example, because of the absence of major naval opponents at the present time, major naval operations to destroy the enemy fleet at sea are not likely to occur in the near future. Nevertheless, it would be very wrong for theoreticians not to pay attention to that type of major naval operation. Likewise, a blue-water navy, while focusing on major naval operations in littoral waters, should not neglect the possibility of a fighting fleet or encounters on the open ocean.

The current highly unsatisfactory situation might continue without adverse consequences for the U.S. Navy and other Western navies as long as they do not face a peer competitor who can ensure superiority not through the number of platforms and advanced weapons but by thinking and acting operationally instead of tactically.

Appendix: Major Naval Operations in World War II

This list is not all-inclusive.

Fleet-versus-Fleet Operations

3 July 1940. Attack by British Force H (1 battle cruiser, 2 battleships, 1 aircraft carrier, 2 light cruisers, 11 destroyers) on a part of the French fleet in Mers-el-Kebir (near Oran) (CATAPULT).

6–10 July 1940. The Italian operation in the central Mediterranean (leads to the battle off Punta Stilo, Calabria).

11–12 November 1940. British carrier attack on the Italian fleet in Tarent (Taranto) (JUDGMENT).

24–29 November 1940. Battle of Cape Teulada between the Italian and the British fleets (COLLAR).

27–29 March 1941. Battle of Matapan (near Mani Peninsula, southern Greece).

18–27 May 1941. Attempt by the battleship *Bismarck* and heavy cruiser *Prinz Eugen* to attack Allied convoys in the central Atlantic (RHEINUEBUNG).

22–26 August 1941. British Force H (1 battleship, 1 aircraft carrier, 1 cruiser, 5 destroyers) mines the approaches to the Italian naval base at Leghorn (Livorno) and conducts a raid on the Italian airfield at Tempio (northern Sardinia) (MINCEMEAT).

25 November–22 December 1941. The Japanese attack on Pearl Harbor (HAWAII) (6 aircraft carriers and 423 aircraft, 2 battleships, 2 heavy cruisers, 1 light cruiser, 11 destroyers, 3 submarines, 8 tankers and supply ships).

12–14 February 1942. The escape of the two German battle cruisers *(Scharnhorst, Gneisenau)* and one heavy cruiser *(Prinz Eugen)* with their escorts from Brest, France, through the English Channel to Wilhelmshaven (CERBERUS).

26 March–11 April 1942. The Japanese carrier raid on Ceylon (5 aircraft carriers, 2 battleships, 2 heavy cruisers, 1 light cruiser, 8 destroyers).

3–8 May 1942. Battle of the Coral Sea (Japanese Operation MO).

25 May–7 June 1942. Battle of Midway (Japanese Operation MI).

18–22 June 1944. Battle of the Philippine Sea.

17–29 October 1944. The Battle for Leyte.

Fleet-versus-Shore Operations

Amphibious Landing Operations

6–15 December 1941. Landing operation by the Japanese South Philippines Force from Palau.

7–12 December 1941. Landings by the Japanese North Philippines Force.

8–12 December 1941. Japanese landing in Malaya.

16–20 December 1941. Japanese landing on Mindanao.

17–24 December 1941. Japanese landing in the Lingayen Gulf and Lamon Bay (Luzon).

14–25 January 1942. Japanese landings at Rabaul (New Britain) and Kavieng (New Ireland).

20–26 January 1942. Japanese landings at Pattani and Singora.

25 February–9 March 1942. Japanese invasion of Java.

8–12 March 1942. Japanese landing in northern Sumatra.

31 July–9 August 1942. Allied landing on Guadalcanal (WATCHTOWER).

11 August 1942. Allied landing at Dieppe, France (JUBILEE).

8–10 November 1942. Allied landing in North Africa (TORCH).

29 June–13 July 1943. Allied landings on New Georgia (TOENAILS).

10 July–17 August 1943. Allied landing on Sicily (HUSKY).

31 August–3 September 1943. Allied landing in Calabria (BAYTOWN).

9–16 September 1943. Allied landing at Salerno (AVALANCHE).

19–28 November 1943. Allied landing on the Gilbert Islands (GALVANIC).

24–27 December 1943. Allied landing on Cape Gloucester, New Britain (DEXTERITY).

2–23 January 1944. Allied landings near Saidor (New Guinea).

21–29 January 1944. Allied landings at Anzio/Nettuno (SHINGLE).

31 January–7 February 1944. Allied landing on the Kwajalein Atoll (FLINTLOCK).

17–23 February 1944. Allied landing on Eniwetok (CATCHPOLE).

19 February 1944. Allied invasion of the Admiralty Islands (BREWER).

22–28 April 1944. Allied landings at Hollandia (RECKLESS) and at Aitape (PERSECUTION).

6 June 1944. Allied landing in Normandy (NEPTUNE).

14 June–9 July 1944. Allied landing on Saipan (FORAGER).

21 July–10 August 1944. Allied landing on Guam (FORAGER).

24 July–1 August 1944. Allied landing on Tinian (FORAGER).

27–31 July 1944. Allied landing east of Cape Sansapor (GLOBE TROTTER).

15 August 1944. Allied landing in southern France (DRAGOON).

15 September 1944. Allied landing on Morotai (TRADEWIND).

15 September–23 October 1944. U.S. amphibious landing on Palau (STALEMATE II).

24 September–31 October 1944. British occupation of the Aegean Islands and the Greek mainland (ACCOLADE).

9–17 January 1945. Allied landing in the Lingayen Gulf (MIKE I).

16 January–4 February 1945. British landing off Arakan (Burma) (MATADOR).

19 February–16 March 1945. Allied landing on Iwo Jima (DETACHMENT).

28 February 1945. Allied landing on Palawan (Philippines) (VICTOR III).

10 March–9 April 1945. Landing near Zamboanga (Mindanao) (VICTOR IV).

18–29 March 1945. Allied landing on Panay (VICTOR I).

26 March–26 April 1945. Allied amphibious landing on Cebu (Philippines) (VICTOR II).

1 April–23 June 1945. Allied landing on Okinawa (ICEBERG).

17–22 April 1945. Allied amphibious landing on the coast of Moro Gulf, western Mindanao (Philippines) (VICTOR V).

27 April–1 May 1945. Allied amphibious landing on Tarakan (Borneo) (OBOE I).

7–20 June 1945. U.S. amphibious landing in Brunei Bay (Borneo) (OBOE VI).

15 June–9 July 1945. U.S. amphibious landing at Balikpapan (Borneo) (OBOE II).

Operations versus Enemy Shore

15–19 September 1940. British attack on Benghazi, Libya, with 1 battleship, 1 aircraft carrier, 1 cruiser, 2 antiaircraft cruisers, and 7 destroyers.

23–25 September 1940. British naval attack on Dakar to prepare a landing by the Free French troops (MENACE).

16–24 December 1940. British force composed of 2 battleships, 2 aircraft carriers, 2 cruisers, and 11 destroyers attacks Italian airfields on Rhodes and Stampalia.

31 January–4 February 1941. Unsuccessful operation by the British Force H (1 battle cruiser, battleship, aircraft carrier, and light cruiser each, and 10 destroyers) against the dam on Lake Tirso, Sardinia. The planned attack on Genoa is canceled due to bad weather.

22 July–4 August 1941. British carrier raid on Kirkenes and Petsamo with Force P (2 aircraft carriers, 2 heavy cruisers, 4 destroyers).

15–24 February 1942. The Japanese carrier raids on Port Darwin, Timor, and Bali.

23 March–6 April 1944. U.S. carrier raids on Palau, Yap, and Woleai (DESECRATE).

6–27 May 1944. British carrier raids on Soerabaja (TRANSOM).

11–17 June 1944. U.S. carrier raids on the Mariana and Vulcan islands in support of Operation FORAGER.

4–21 July 1944. U.S. carrier raids on Iwo Jima and Chichi Jima.

28 August–24 September 1944. Preparatory strikes in support of landing on Palau-Morotai. U.S. carrier raids on Iwo Jima and Chichi Jima, the Visayas (central Philippines), Mindanao, and Luzon.

6–17 October 1944. U.S. Third Fleet carrier raid on Ryukyus, Formosa, and Luzon.

21–30 January 1945. British carrier raids on Sumatra (MERIDIAN).

14–23 March 1945. TF 58 carrier raid on Japan.

24–30 July 1945. U.S. and British carrier raids against Japan.

Operations versus Enemy Merchant Shipping

18–20 February 1940. German attack on convoy traffic between England and Scandinavia (2 battle cruisers, 1 heavy cruiser, 2 destroyers, and 6 U-boats) (NORDMARK).

4–10 June 1940. The German fleet (2 battle cruisers, 1 heavy cruiser, 4 destroyers) attacks British evacuation transports in the area west of Harstad (JUNO).

9–22 September 1940. German U-boats against HX.72 convoy in the North Atlantic.

9–20 October 1940. German U-boats against SC.7 convoy in the North Atlantic.

8–12 February 1941. A German heavy cruiser and U-boats against HG.53 and SLS.64 in the North Atlantic.

19–22 February 1941. German U-boats against OB.287 convoy in the North Atlantic.

22–24 February 1941. German U-boats against OB.288 convoy west of Ireland.

23 February–3 March 1941. German U-boats against OB.289 convoy in the North Atlantic.

24–26 February 1941. German U-boats against OB.290 convoy in the North Atlantic.

7–10 March 1941. Two German battle cruisers and U-boats against SL.67 convoy in the central Atlantic.

10–17 March 1941. German U-boats against HX.112 convoy (41 ships) in the North Atlantic.

15–21 March 1941. German U-boats against SL.68 convoy off West Africa.

17–29 March 1941. German U-boats against HX.115 convoy (BERLIN).

19–24 March 1941. German U-boats and Italian submarines against OG.56 convoy west of Ireland.

25 March–5 April 1941. German U-boats against OB.302, HX.115, and SC.26 in the North Atlantic.

7–30 April 1941. German U-boats and Italian submarines against HX.121 convoy in the North Atlantic.

1–11 May 1941. German U-boats against HX.122, HX.123, and OB.318 in the North Atlantic.

20–29 June 1941. German U-boats against HX.133 and OB.336 in the North Atlantic.

27 June–4 July 1941. German U-boats against SL.76 convoy in the central Atlantic.

29 June–15 July 1941. German U-boats against OG.66 convoy in the North Atlantic.

18–30 July 1941. German and Italian submarines against HG.67, HG.68, OG.68, OG.69, and SL.80 in the North Atlantic.

5–16 August 1941. German U-boats against HX.142, HX.143, and SC.39 in the North Atlantic.

6–16 August 1941. German U-boats and the Italian submarines against HG.69 convoy in the North Atlantic.

10–23 August 1941. German U-boats against OG.71 convoy in the North Atlantic.

18 August–8 September 1941. The Italian submarines against Allied convoys HG.71 and HG.72 in the North Atlantic.

22–30 August 1941. The German U-boats against Allied convoy OS.4 in North Channel, Ireland.

28 August–14 September 1941. German U-boats against ON.10S, HX.147, ON.12S, SC.42, SC.41, HX.148, SC.43, and EG.18 in the North Atlantic.

1–18 September 1941. German U-boats against SL.84, OG.73, HG.72, and SC.42 in the northern Atlantic.

15 September–2 October 1941. German U-boats against SC.44 (56 ships) and ONS.19 in the North Atlantic.

15 September–6 October 1941. German U-boats against HX.150 (50 ships), SC.45 (60 ships), and ON.18 (25 ships) in the North Atlantic.

18–28 September 1941. German U-boats and Italian submarines on the Gibraltar route against OG.74 (27 ships) and HG.73 (25 ships).

24 September–1 October 1941. German U-boats against HX.151 (44 ships), ON.20 (51 ships), and SC.46 (61 ships) in the North Atlantic.

11–26 October 1941. German U-boats against HX.154 (47 ships), SC.49, and OB.26 (32 ships) in the North Atlantic.

17–29 October 1941. German U-boats against HG.75 convoy in the North Atlantic.

18 October–2 November 1941. German U-boats against SL.89, HX.155 (59 ships), ONS.27 (62 ships), and ON.28 (40 ships) in the North Atlantic.

24 October–4 November 1941. German U-boats against HX.156 (44 ships), SC.51 (37 ships), ONS.29 (34 ships), and OS.10 in the North Atlantic.

1–8 November 1941. German U-boats against HX.157 (44 ships), ON.28, SC.52, ONS.29, ON.31 (33 ships), and ONS.32 (49 ships) in the North Atlantic.

5–16 November 1941. German U-boats against HX.158 (40 ships), SC.53, ONS.33 (54 ships), and ON.34 (46 ships) in the North Atlantic.

11 November–2 December 1941. German U-boats against HX.159 (33 ships), OS.15, SC.54, OG.77, HX.160 (62 ships), ONS.36 (39 ships), and OS.12 in the North Atlantic.

21 November–30 December 1941. German U-boats against HX.161, ON.43, SC.56, ONS.44, HX.163, ONS.48, SC.58, ON.45, and HX.164 in the North Atlantic.

14–23 December 1941. German U-boats against HG.76 convoy in the North Atlantic.

January–July 1942. German U-boats against shipping off U.S. east coast (PAUKENSCHLAG).

26 April–12 May 1942. German operation against Allied convoys PQ.15 and QP.11 in the Arctic.

25 May–1 June 1942. German attack on Allied convoys QP.12 (15 ships) and PQ.16 (35 ships) in the Arctic (ROESSELSPRUNG–KNIGHT'S GAMBIT).

20 June–28 July 1942. German operations against Allied convoys QP.13 (12 ships) and PQ.17 (36 ships) in the Arctic.

July 1942. German U-boats against OS.33 and OS.34 in the North Atlantic.

August 1942. German U-boats against ON.115 and SC.94 in the North Atlantic.

August 1942. German U-boats against ONS.122 convoy in the North Atlantic.

August 1942. German U-boats against SL.118 and SL.119 in the North Atlantic.

August–September 1942. German U-boats against SC.97 and ON.127 in the North Atlantic.

11–30 August 1942. German attack on the Soviet shipping on the Siberian sea route (WUNDERLAND).

12–18 September 1942. German operation against Allied supply convoy PQ.18 in the Arctic.

September 1942. German U-boats against SC.99 and ON.129 in the North Atlantic.

17–25 September 1942. German U-boats against SC.100 convoy in the North Atlantic.

22–29 September 1942. German U-boats against ON.131 and RB.1 in the North Atlantic.

30 September–6 October 1942. German U-boats against HX.209 convoy in the North Atlantic.

8–16 October 1942. German U-boats against ONS.136 and SC.104 in the North Atlantic.

24 October–7 November 1942. German U-boats against SC.107 and ON.143 in the North Atlantic.

13–22 December 1942. The German operation against HX.218, ONS.152, and ON.153 in the North Atlantic.

14–25 February 1943. German U-boats against ON.166 and ONS.167 in the North Atlantic.

27 February–11 March 1943. German U-boats against SC.121 convoy in the North Atlantic.

7–14 March 1943. German U-boats against ON.170 and HX.228 convoys in the North Atlantic.

15–20 March 1943. German U-boats against SC.122 and HX.229 convoys in the North Atlantic.

22–30 March 1943. German U-boats against HX.230 convoy in the North Atlantic.

7–13 April 1943. German U-boats against ON.176 and HX.232 convoys in the North Atlantic.

14–25 April 1943. German U-boats against HX.233 and HX.234 convoys in the North Atlantic.

26 April–3 May 1943. German U-boats against ONS.5 and SC.128 convoys in the North Atlantic.

16 June–10 July 1943. German U-boats against GU.S and UGS convoys in the central Atlantic.

1 August–3 September 1943. German operations on the Siberian sea route (WUNDERLAND II).

2–23 September 1943. German U-boats against ON.202 and ONS.18 convoys in the North Atlantic.

11–16 November 1943. German U-boats against HX.264, ON.210, and ONS.22 convoys in the North Atlantic.

10–26 April 1944. Soviet combined Operation RC-3 against German convoy traffic off Norwegian polar coast.

11–29 May 1944. Soviet operation RV-4 against German traffic off Norway's northern coast.

16–31 August 1944. Soviet operation against German traffic off Norway's northern coast.

10–28 September 1944. Soviet operation against German traffic off Norway's northern coast.

11–21 March 1945. German operation against JW.65 convoy in the Arctic.

23–31 March 1945. German operation against RA.65 convoy in the Arctic.

29 April–2 May 1945. German operation against JW.66 convoy (last convoy operation of World War II).

Operations to Protect Friendly Merchant Shipping

10–14 July 1940. Convoy from Malta to Alexandria (MF1); escort provided by four battleships, one large aircraft carrier, seven cruisers, and twenty-one destroyers.

29 August–2 September 1940. Convoy from Alexandria to Malta (MF2); escort provided by three cruisers and four destroyers.

8–11 October 1940. Convoy from Alexandria to Malta (MF3); escort provided by three battleships, two large aircraft carriers, six cruisers, and seventeen destroyers.

4–11 November 1940. Convoy from Alexandria to Malta (MW3); escort provided by four battleships, two antiair cruisers, four destroyers, and one old minesweeper.

10–13 November 1940. Convoy from Malta to Alexandria (ME3); escort provided by one battleship, one cruiser, three destroyers, and one monitor.

15–20 November 1940. Force H (1 battle cruiser, 1 aircraft carrier, 2 cruisers, 8 destroyers) accompanies 1 aircraft carrier to the southwest of Sardinia to fly off 12 Hurricanes and 2 Sea Skuas to Malta (WHITE).

24–30 November 1940. Dual convoy operation (COLLAR) launched from the western and eastern Mediterranean; one convoy of two ships sails from Gibraltar and another convoy of four ships from Alexandria to bring supplies to Malta. The westbound convoy, escorted by one battleship, three cruisers, and five destroyers, reaches Malta on 27 November.

26–29 November 1940. Convoy (five ships) from Malta to Alexandria/Port Said (ME4); escorted by one cruiser and three destroyers. Distant cover and support provided by one battleship, one aircraft carrier, one cruiser, and five destroyers.

16–20 December 1940. Convoy (six ships) from Port Said/Alexandria to Malta (MW 5B); escort provided by two battleships, one aircraft carrier, two cruisers, and eleven destroyers.

16–24 December 1940. Resupply convoy to Malta (MC.2).

6–13 January 1941. Convoys to Malta and Piraeus, Greece (EXCESS).

5 March–2 April 1941. Transport of four British divisions (58,000 troops) from Alexandria to Greece (LUSTRE).

5–12 May 1941. Allied resupply convoy from Gibraltar to Malta (TIGER).

11–14 May 1941. The Italian-German convoy to Libya.

19–21 May 1941. The Italian-German convoy from Palermo to Libya.

5–7 June 1941. The British Force H (1 battle cruiser, 2 aircraft carriers, and 6 destroyers) sails out from Gibraltar to the western Mediterranean to fly off 36 fighters to Malta (ROCKET).

21–27 July 1941. Allied supply convoy from Gibraltar to Malta (SUBSTANCE).

21–30 July 1940. Allied convoy from Alexandria/Port Said to the Aegean (AS.2 return convoy) (AN.2).

19–23 August 1941. The Italian convoy to Tripoli, Libya.

21–31 August 1941. The first Allied convoy from Hvalfjord, Iceland, to Arkhangel'sk (DERVISH).

29 August–4 September 1941. A large Italian convoy from Naples to Tripoli.

24–30 September 1941. Convoy of 9 freighters (85,000 tons of supplies) from Gibraltar to Malta escorted by 1 carrier, 3 battleships, 5 cruisers, and 18 destroyers (HALBERD).

8–15 October 1941. The Italian convoy Giulia from Naples to Libya.

10 November–27 December 1941. The convoy WS.24 sails out from Halifax, Nova Scotia, to the Near East with 20,000 troops on board.

16 December 1941. The Italian convoy operation M.42 to Libya.

3–6 January 1942. The Italian supply operation M.43 to Libya.

5–9 January 1942. British supply to Malta (MF.2).

22–25 January 1942. The Italian supply to Tripoli, Libya (Operation T.18).

24–28 January 1942. British supply to Malta (MF.4).

12–16 February 1942. British supply to Malta (MF.5).

19 February–2 March 1942. Troop convoy AT–12 with 13 ships and 14,700 men escorted by TF.32 (1 battleship, 1 heavy cruiser, 11 destroyers).

21–23 February 1942. Two Italian convoys from Messina and Corfu to Tripoli (K.7).

7–18 March 1942. The Italian supply convoy to North Africa (V.5).

20–26 March 1942. British supply convoy MW.10 for Malta.

12–16 June 1942. Dual convoy operation (HARPOON/VIGOROUS) from Gibraltar and Alexandria to supply Malta.

10–15 August 1942. Supply convoy (14 ships) from Gibraltar to Malta (PEDESTAL).

12 November–2 December 1942. Allied convoy (45 ships) from U.S. to Casablanca escorted by TF.37 (6 destroyers, 2 tankers).

12–30 December 1942. Allied convoy (45 ships) from New York to North Africa escorted by TF.35 (1 battleship, 1 heavy cruiser, 8 destroyers, 2 tankers).

17 January–8 February 1943. Allied convoy operations JW.52/RA.52 in the Arctic.

4–27 February 1943. Convoy with 30,000 men of the 9th Australian Division sails from Suez to Sydney and Melbourne (PAMPHLET).

15 February–14 March 1943. Allied convoy JW.53 (28 ships) in the Arctic.

27 March–28 April 1943. The Axis supply convoy operation to Tunisia.

12 December 1943–8 January 1944. Convoy operations in the Arctic; convoys JW.55A, JW.55B, and RA.55A.

20 February–10 March 1944. Convoy operations JW.57/RA.57 in the Arctic.

27 March–5 April 1944. Convoy operations JW.58/RA.58 in the Arctic.

30 March–1 April 1944. Convoy UGS.36 (72 ships) and 18 LSTs escorted by TF.64 (4 destroyers, 7 escort destroyers) and the British 37th Escort Group (1 antiaircraft cruiser, 1 Dutch frigate, 1 corvette, 1 minesweeper) transit the Strait of Gibraltar to the Mediterranean.

19–20 April 1944. Allied convoy UGS.38 (87 ships) escorted by TF.66 transits the Strait of Gibraltar to the Mediterranean.

3–5 May 1944. Allied convoy GUS.38 (107 ships) escorted by TF.66 transits the Strait of Gibraltar from the east.

9–11 May 1944. Allied convoy UGS.40 (65 ships) escorted by TF.60 transits the Strait of Gibraltar.

15 August–6 September 1944. Convoy operations JW.59/RA.59A in the Arctic.

15 September–5 October 1944. Convoy operations JW.60/RA.60 in the Arctic.

20–29 October 1944. Soviet convoy operation AB.15 in the Arctic.

20 October–10 November 1944. Convoy operations JW.61/RA.61 in the Arctic.

27 November–14 December 1944. Convoy operations JW.62/RA.62 in the Arctic.

3 December 1944–2 January 1945. Convoy operations JW.63/RA.63 in the Arctic.

3–13 February 1945. Convoy operation JW.64 to Murmansk with 26 ships and escorted by 2 escort carriers, 1 cruiser, and 17 escort vessels (including 6 destroyers).

14–28 February 1945. Convoy operation RA.64 in the Arctic.

11–21 March 1945. Convoy operation JW.65 in the Arctic.

16–25 April 1945. Convoy operation JW.66 in the Arctic.

12–31 May 1945. Last convoy pair, JW.67/RA.67, in the Arctic.

Special-Purpose Major Naval Operations

17 October 1940. 12 Hurricane fighters flown from carrier *Argus* to Malta (only 4 survived) (WHITE).

2–5 April 1941. Two British carriers fly off 12 fighters from the western Mediterranean to Malta (WINCH).

4–28 April 1941. Force H (1 battle cruiser, 1 aircraft carrier, 2 cruisers, 5 destroyers) flies off 24 Hurricanes to Malta (DUNLOP).

19–22 May 1941. Force H (1 battle cruiser, 2 aircraft carriers, 1 heavy cruiser, 6 destroyers) south of Sardinia flies off 48 fighters to Malta (SLICE) (23 hurricanes reached Malta).

14 June 1961. 48 Hurricanes fly off from 2 carriers to Malta (TRACER) (45 reach Malta).

8–14 September 1941. British force composed of 1 aircraft carrier, 1 cruiser, and 6 destroyers sails from Gibraltar to an area south of Balearics from which 14 fighters are flown to Malta. In a second sortie 1 battleship and 1 destroyer take part. On 13 September 2 aircraft carriers fly off another 45 fighters to Malta (STATUS).

10–14 November 1941. British Force H (2 aircraft carriers, 1 battleship, 1 light cruiser, 7 destroyers) sails out from Gibraltar to the western Mediterranean and flies off 37 fighter aircraft to Malta (PERPETUAL).

20–30 March 1942. British Force H (1 battleship, 2 carriers, 1 cruiser, 9 destroyers) sets out from Gibraltar to make new attempt to fly off aircraft into Malta. But the force has to break off the attempt and return on 23 March. The operation is repeated from 27 to 30 March with the same force, and 16 Spitfire fighters reach Malta.

14–26 April 1942. Anglo-American Force W (1 battle cruiser, 1 aircraft carrier, 6 destroyers) sails out of Greenock on 14 April to Gibraltar, where it is reinforced with 2 British cruisers and flies off 47 Spitfire fighters for Malta on 20 April; 46 of them arrive (CALENDAR).

3–15 May 1942. The Anglo-American Force W (1 British battle cruiser, 1 U.S. carrier, 1 British cruiser, 2 U.S. destroyers) sets out from Scapa Flow on 3 May. During the night of 7–8 May this force is joined by 1 British carrier and 9 destroyers from Gibraltar. On 9 May, a total of 64 Spitfires are flown from the carriers to Malta (BOWERY).

17–20 May 1942. British Force H sets out from Gibraltar with 2 carriers, 1 cruiser, and 7 destroyers on 17 May. The following day it flies off 17 Spitfire fighters and 6 Albacore torpedo planes to Malta (the latter are forced to return).

2–4 June 1942. British Force H (1 aircraft carrier, 1 cruiser, 5 destroyers) sails out from Gibraltar and flies off 31 Spitfire fighters on 3 June south of Balearics to Malta (STYLE).

14–19 July 1942. Part of British Force H (1 aircraft carrier, 2 cruisers, 5 destroyers) leaves Gibraltar to fly off 31 Spitfire fighters to Malta (PINPOINT).

20–22 July 1942. British Force H (1 aircraft carrier, 2 cruisers, 5 destroyers) flies off 28 Spitfire fighters for Malta (INSECT).

16–18 August 1942. British Force H (1 aircraft carrier, 1 cruiser, 12 destroyers) sets out from Gibraltar to the area south of Balearics and flies off 32 Spitfire fighters to Malta (BARITONE).

List of Abbreviations and Acronyms

A

AKA	attack cargo ship
AOA	amphibious objective area
APA	attack transport
ASCM	antiship cruise missile
ASW	antisubmarine warfare

C

C2W	command and control warfare
C4	command, control, communications, and computers
CFMCC	combined force maritime component commander
CINCPAC	Commander in Chief, Pacific Fleet
CINCPOA	Commander in Chief, Pacific Ocean Area
COMJTF	commander, maritime joint task force
CONOPS	concept of operations
CSG	carrier strike group

E

EC	enemy capability
ECOA	enemy course of action (or intention)
ESG	expeditionary strike group

I

IJN	Imperial Japanese navy
IO	information operations

J

JFMCC	joint force maritime component commander

L

LCT	landing craft, tank
LSD	landing ship, dock
LST	landing ship, tank

O

OKW	Oberkommando der Wehrmacht (high command of the German armed forces in World War II)
OOB	order of battle
OPLAN	operation plan
OPORD	operation order
OPSEC	operations security

R

RCT	regimental combat team
ROE	rule(s) of engagement

S

SAM	surface-to-air missile
SSBN	nuclear-powered ballistic-missile submarine
SSN	nuclear-powered attack submarine

T

TAP	transport
TF	task force
TG	task group
TLAM	theater land-attack missile

U

UN	United Nations

W

WMD	weapon(s) of mass destruction

About the Author

Dr. Milan Vego has been a professor in the Joint Military Operations Department at the U.S. Naval War College, Newport, Rhode Island, since August 1991. A native of Bosnia and Herzegovina, he obtained political asylum in the United States in 1976. Dr. Vego has been an adjunct professor at the Defense Intelligence College (1984–91) and a senior fellow at the Center for Naval Analyses in Alexandria, Virginia (1985–87), and he has taught at the former Soviet Army Studies Office, Fort Leavenworth, Kansas (1987–89). He earned a BA (1970) and an MA in modern history (1973) at the University of Belgrade and his PhD in European history from the George Washington University (1981). He holds a master mariner license. Dr. Vego has published six books, including the textbook *Operational Warfare* (2001), and numerous articles in various professional journals. His most recent book, *The Battle for Leyte, 1944: Allied and Japanese Plans, Preparations, and Execution,* was published by the Naval Institute Press in March 2006.

The Newport Papers

Perspectives on Maritime Strategy: Essays from the Americas, edited by Paul D. Taylor (no. 31, August 2008).

U.S. Naval Strategy in the 1970s: Selected Documents, edited by John B. Hattendorf (no. 30 September 2007).

Shaping the Security Environment, edited by Derek S. Reveron (no. 29, September 2007).

Waves of Hope: The U.S. Navy's Response to the Tsunami in Northern Indonesia, by Bruce A. Elleman (no. 28, February 2007).

U.S. Naval Strategy in the 1990s: Selected Documents, edited by John B. Hattendorf (no. 27, September 2006).

Reposturing the Force: U.S. Overseas Presence in the Twenty-first Century, edited by Carnes Lord (no. 26, February 2006).

The Regulation of International Coercion: Legal Authorities and Political Constraints, by James P. Terry (no. 25, October 2005).

Naval Power in the Twenty-first Century: A Naval War College Review *Reader,* edited by Peter Dombrowski (no. 24, July 2005).

The Atlantic Crises: Britain, Europe, and Parting from the United States, by William Hopkinson (no. 23, May 2005).

China's Nuclear Force Modernization, edited by Lyle J. Goldstein with Andrew S. Erickson (no. 22, April 2005).

Latin American Security Challenges: A Collaborative Inquiry from North and South, edited by Paul D. Taylor (no. 21, 2004).

Global War Game: Second Series, 1984–1988, by Robert Gile (no. 20, 2004).

The Evolution of the U.S. Navy's Maritime Strategy, 1977–1986, by John Hattendorf (no. 19, 2004).

Military Transformation and the Defense Industry after Next: The Defense Industrial Implications of Network-Centric Warfare, by Peter J. Dombrowski, Eugene Gholz, and Andrew L. Ross (no. 18, 2003).

The Limits of Transformation: Officer Attitudes toward the Revolution in Military Affairs, by Thomas G. Mahnken and James R. FitzSimonds (no. 17, 2003).

The Third Battle: Innovation in the U.S. Navy's Silent Cold War Struggle with Soviet Submarines, by Owen R. Cote, Jr. (no. 16, 2003).

International Law and Naval War: The Effect of Marine Safety and Pollution Conventions during International Armed Conflict, by Dr. Sonja Ann Jozef Boelaert-Suominen (no. 15, December 2000).

Theater Ballistic Missile Defense from the Sea: Issues for the Maritime Component Commander, by Commander Charles C. Swicker, U.S. Navy (no. 14, August 1998).

Sailing New Seas, by Admiral J. Paul Reason, U.S. Navy, with David G. Freymann (no. 13, March 1998).

What Color Helmet? Reforming Security Council Peacekeeping Mandates, by Myron H. Nordquist (no. 12, August 1997).

The International Legal Ramifications of United States Counter-Proliferation Strategy: Problems and Prospects, by Frank Gibson Goldman (no. 11, April 1997).

Chaos Theory: The Essentials for Military Applications, by Major Glenn E. James, U.S. Air Force (no. 10, October 1996).

A Doctrine Reader: The Navies of the United States, Great Britain, France, Italy, and Spain, by James J. Tritten and Vice Admiral Luigi Donolo, Italian Navy (Retired) (no. 9, December 1995).

Physics and Metaphysics of Deterrence: The British Approach, by Myron A. Greenberg (no. 8, December 1994).

Mission in the East: The Building of an Army in a Democracy in the New German States, by Colonel Mark E. Victorson, U.S. Army (no. 7, June 1994).

The Burden of Trafalgar: Decisive Battle and Naval Strategic Expectations on the Eve of the First World War, by Jan S. Breemer (no. 6, October 1993).

Beyond Mahan: A Proposal for a U.S. Naval Strategy in the Twenty-First Century, by Colonel Gary W. Anderson, U.S. Marine Corps (no. 5, August 1993).

Global War Game: The First Five Years, by Bud Hay and Bob Gile (no. 4, June 1993).

The "New" Law of the Sea and the Law of Armed Conflict at Sea, by Horace B. Robertson, Jr. (no. 3, October 1992).

Toward a Pax Universalis: A Historical Critique of the National Military Strategy for the 1990s, by Lieutenant Colonel Gary W. Anderson, U.S. Marine Corps (no. 2, April 1992).

"Are We Beasts?" Churchill and the Moral Question of World War II "Area Bombing," by Christopher C. Harmon (no. 1, December 1991).

Newport Papers from 16 on are available online (Acrobat required) at www.nwc.navy.mil/press/newportpapers/newportpapers.aspx.

Made in the USA
Monee, IL
04 November 2026

67381291R00085